Methods for Fragments Screening Using Surface Plasmon Resonance

Sameer Mahmood Zaheer •
Ramachandraiah Gosu
Editors

Methods for Fragments Screening Using Surface Plasmon Resonance

Springer

Editors
Sameer Mahmood Zaheer
Structural Biology Division, Discovery Biology
Jubilant Biosys Ltd.
Bengaluru, Karnataka, India

Ramachandraiah Gosu
Structural Biology Division, Discovery Biology
Jubilant Biosys Ltd.
Bengaluru, Karnataka, India

ISBN 978-981-16-1535-1 ISBN 978-981-16-1536-8 (eBook)
https://doi.org/10.1007/978-981-16-1536-8

This Springer imprint is published by the registered company Springer Nature Singapore Pte Ltd.
The registered company address is: 152 Beach Road, #21-01/04 Gateway East, Singapore 189721, Singapore

This book is dedicated to the people, who believe in Fundamentals of Science

Preface

This book was motivated by the aspiration we had to support the screening of fragments using SPR-based HTS in drug discovery process. We believe that this book will create a great deal in addressing step-by-step protocol from identification of fragments from million libraries, production of SPR compatible protein to identification of fragment hits. It has been a challenge to bring together the different sections in perspective to achieve the objective of identifying a true fragment hit. The target of this book is not only the Pharma and Biotech industries, but also the academic institutions particularly the research centers where the screening of fragments are routinely carried out. This book is written keeping these audiences in mind.

The primary goal of this book is to provide scientists a working protocol which has been optimized and validated at Jubilant and will assist these researchers. Thereby, the probability to reduce the process of fragment screening is emphasized here. Also it will benefit the whole drug discovery community.

The procedures are discussed in each chapter in detail with notes and respective troubleshooting giving a clear understanding, the important modules responsible for the success of experiments. As HTS screening with SPR is gaining prominence in drug discovery, it is imperative for researchers to know and learn about this platform and prioritize their preference.

Bengaluru, India Sameer Mahmood Zaheer
Bengaluru, India Ramachandraiah Gosu

Acknowledgments

We thank each and every author who have contributed in this book. We are grateful to Dr. Saravanakumar Dhakshinamoorthy, Vice President, Discovery Biology and Dr. Takeshi Yura, Site Head of Jubilant Biosys Ltd. for giving us the opportunity and permission to write this book. We would like to appreciate Mr. Ravi Syam Madhira, Director, Intellectual Property, Jubilant Biosys Ltd. for helping in understanding the legality while writing this book. Also we would like to extend our thanks to Sartorius team members Dr. Susheelendra Vaidya and Mr. Ishan Subudhi, for providing the basic customized SPR figures. Special thanks to Ms. Aswathy Pillai for proofreading the entire book chapters.

Contents

Editors and Contributors

About the Editors

Sameer Mahmood Zaheer is Group Leader at the Jubilant Biosys Ltd., Bengaluru, India. He has earlier served as Associate Principal Scientist in the Department of invitro biology at GVK Bioscience Ltd., Hyderabad, India, and Principal Scientist I in the Department of Molecular and Cell Biology, Advinus Therapeutics Ltd., Pune. He also did postdoctoral training in the Department of Stem Cell and Molecular Biology at Joslin Diabetes Center (Harvard University), Boston, USA. His research interest is in the high throughput screening of proteins using robotics with specialization in protein-drug (Fragments and Small molecules) interactions. He has been accolade with several awards in Stem cell research, Jackson Laboratory (USA), Asia ARVO (Singapore), Research Achievement Award (USA), CEO award for Excellence in Project Management (India).

He has published several research articles in peer-reviewed international journals and co-authored a book chapter.

Ramachandraiah Gosu is Associate Director, Discovery Biology & Head of Structural Biology Department at Jubilant Biosys Ltd., Bengaluru. His research interests are protein engineering, macromolecular X-ray crystallography and methods for biophysical screening. Before joining Jubilant Biosys, he worked as a post-doctoral fellow at the University of Minnesota, Minneapolis in the Structural Genomics Project—Pathogenic Island Proteins from Staphylococcus aureus. He has obtained his Ph.D. in, Structural Biology from Molecular Biophysics Unit at the Indian Institute of Science, Bangalore, India. He has co-authored several articles in peer-reviewed international journals.

Contributors

Rashmi Rekha Devi Structural Biology Division, Discovery Biology, Jubilant Biosys Ltd., Bengaluru, Karnataka, India

Ramachandraiah Gosu Structural Biology Division, Discovery Biology, Jubilant Biosys Ltd., Bengaluru, Karnataka, India

Saravanan Kandan Structural Biology Division, Discovery Biology, Jubilant Biosys Ltd., Bengaluru, Karnataka, India

Rajendra Kristam Computational Chemistry Division, Medicinal Chemistry, Jubilant Biosys Ltd., Bengaluru, Karnataka, India

Aswathy Pillai Structural Biology Division, Discovery Biology, Jubilant Biosys Ltd., Bengaluru, Karnataka, India

Srinivasan Swaminathan Structural Biology Division, Discovery Biology, Jubilant Biosys Ltd., Bengaluru, Karnataka, India

V. Swarnakumari Structural Biology Division, Discovery Biology, Jubilant Biosys Ltd., Bengaluru, Karnataka, India

A. L. Theerthan Structural Biology Division, Discovery Biology, Jubilant Biosys Ltd., Bengaluru, Karnataka, India

Dinesh Thiagaraj Structural Biology Division, Discovery Biology, Jubilant Biosys Ltd., Bengaluru, Karnataka, India

Krishnakumar Vaithilingam Structural Biology Division, Discovery Biology, Jubilant Biosys Ltd., Bengaluru, Karnataka, India

Ponni Vijayan Structural Biology Division, Discovery Biology, Jubilant Biosys Ltd., Bengaluru, Karnataka, India

Sameer Mahmood Zaheer Structural Biology Division, Discovery Biology, Jubilant Biosys Ltd., Bengaluru, Karnataka, India

Abbreviations

AlogP	Octanol/water partition coefficient
Avi	Avidin
°C	Degree Centigrade
CDH	Carboxymethyl Dextran Hydrogel
Da	Dalton
DMSO	Dimethyl Sulfoxide
DSF	Differential Scanning Fluorimetry
DTT	Dithiothreitol
EDC	1-Ethyl-3-(-3-dimethylaminopropyl) carbodiimide
Em	Emission
Ex	Excitation
FC	Flow Channel
Fig	Figure
FP	Fluorescent Polarization
FPLC	Fast protein liquid chromatography
Fsp3	Fraction of sp3 hybridization
ΔG	Change in free energy
g	gravitational force
gm	gram
HAC	Heavy atom count
HBA	Hydrogen Bond Acceptor
HBD	Hydrogen Bond Donor
HEPES	(4-(2-hydroxyethyl)-1-piperazineethanesulfonic acid)
His	Histidine
h	hour
HTS	High-Throughput Screening
IMAC	Immobilized metal affinity chromatography
ITC	Isothermal Titration Calorimetry
IPTG	Isopropyl β-D-1-thiogalactopyranoside
kDa	kilo Dalton
K_D	Equilibrium Dissociation constant
koff	Dissociation rate constant
kon	Association rate constant

L	Liter
LE	Ligand Efficiency
LogS	Aqueous Solubility
MC	Microcalibration
μM	Micromolar
μL	Microliter
mM	Millimolar
mg	Milligram
mL	Milliliter
min	Minute
M	Molar
ng	Nanogram
NHS	*N*-Hydroxysuccinimide
Ni-NTA	Nickel Nitrilotriacetic acid
Ni-IDA	Iminodiacetic acid
nm	Nanometer
NMR	Nuclear Magnetic Resonance
No.	Number
#rings	Number of rings
OD	Optical Density
PAGE	Polyacrylamide Gel Electrophoresis
PAINS	Pan Assay Interference Compounds
%	Percent
PBS	Phosphate-Buffered Saline
pI	Isoelectric point
PSA	Polar Surface Area
REOS	Rapid Elimination of Swill
RFU	Relative Fluorescence Unit
RI	Refractive Index
rpm	Rotations per minute
RT	Room Temperature
RU	Response Unit
SADH	Streptavidin Dextran Hydrogel
SDF	Structure Data File
SDS	Sodium Dodecyl Sulfate
SEC	Size Exclusion Chromatography
TCEP	Tris (2-carboxyethyl)phosphine)
Tm	Melting Temperature
TIR	Total Internal Reflection
TSA	Thermal Shift Assay
U	Unit

Introduction to Surface Plasmon Resonance 1

Ramachandraiah Gosu and Sameer Mahmood Zaheer

Abstract

Several first-in-class drug targets (kinases, dedrogenases, reductases, transferase, etc.) suffer from validation in primary assays due to lack of proper reference molecules. In the absence of reference molecules against any drug targets, the Surface Plasmon Resonance (SPR) based approach will be beneficial for the development of small molecule (combination of various fragments), which may serve as the starting material to synthesize different chemotype series.

Fragment-based screening approach is gaining significance in early stage drug discovery programs. Identifying specific and unique fragment need a highly sensitive platform to differentiate between a genuine hit and false positive hits. SPR from PioneerFE, a label free biosensor technology has a gradient dispersion of analytes (OneStep technology). This OneStep injection's high-resolution concentration gradient assists in more accurate analysis of fragments in contrast to the multi-concentration injections (fixed concentration) required with traditional SPR. We used this technology to develop a High Throughput Screening (HTS) protocol to prioritize high affinity fragments based on dissociation constant (K_D) and ligand efficiency (LE). We screened ~1000 fragments with SPR and positioned them into high, medium, low hits. This is the first-in-class protocol developed using PioneerFE SPR system for fragment screening.

Keywords

Surface plasmon resonance (SPR) · High throughput screening (HTS) · Dissociation constant (K_D) · Ligand efficiency (LE)

R. Gosu (✉) · S. M. Zaheer
Structural Biology Division, Discovery Biology, Jubilant Biosys Ltd., Bengaluru, Karnataka, India
e-mail: ramachandraiah.gosu@jubilanttx.com; Sameer.Mahmood@jubilantbiosys.com

S. M. Zaheer, R. Gosu (eds.), *Methods for Fragments Screening Using Surface Plasmon Resonance*, https://doi.org/10.1007/978-981-16-1536-8_1

1.1 Introduction

In the recent past fragment based drug screening has primed itself as a popular platform for the identification of hits in several drug discovery programs. Major success has been in the detection, characterization, and validation of small molecule fragments (Mw: 100–250 Da) interacting with its binding partner with low affinity (1–10 mM). Several biophysical platforms currently available, like nuclear magnetic resonance (NMR), X-ray crystallography, differential scanning fluorimetry (DSF), and surface plasmon resonance (SPR) have become core technologies in different pharma, biotech, and universities (Table 1.1). Although these biophysical platforms are highly sensitive, fragment screening and low throughput conceive some of them challenging to use on a routine basis [1]. On the other hand, SPR based platforms have similar sensitivity with high throughput mode to provide complete fragment screens on libraries of several thousand fragments (Table 1.2) [2].

Table 1.1 Different biophysical platforms

Platforms	Ligand required	Immobilization of ligand	Analyte required	Screening throughput	Sensitivity
ITC	Low	None	High	Low	nM – mM
NMR	High	None	High	Medium	nM – mM
DSF	Medium	None	High	Medium	uM – mM
FP	Low	None	Intermediate	High	uM
SPR	Low	Yes	Low	High	pM – mM

Table 1.2 SPR based platforms

Manufacturer	Model	Analyte technology	Screening format	Flow cells	Competition assay
Fortebio	PioneerFE	OneStep	96 and 384 well plate	3	Yes – NextStep
Cytiva	Biacore 8K	Fixed concentration	96 and 384 well plate	16	No
Reichert Technologies	Reichert4SPR	Fixed concetration	96 and 384 well plate	4	No
Nicoya	Nicoya Alto	Integrate digital microfluidics	96 and 384 well plate	16	No
Bruker	Sierra SPR-32/24 Pro	Fixed concentration	96 and 384 well plate	8	No
Xantec	SPR-PLUS	Fixed concentration	96 and 384 well plate	2 and 4	No

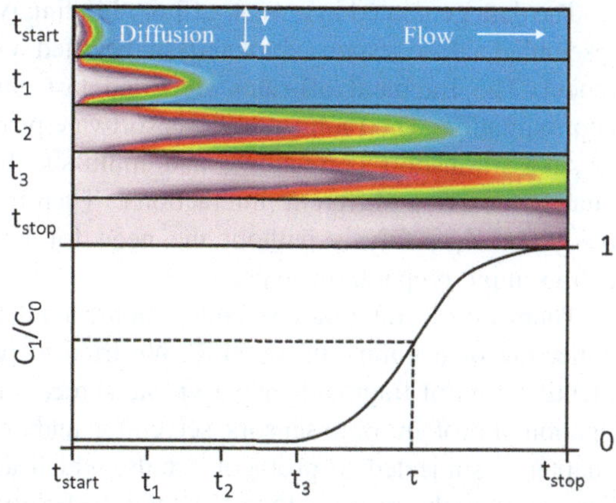

Fig. 1.1 OneStep gradient formatin in the Pioneer injection line (top) and analyte concetration measured at the SPR flow cell (bottom). Blue indicates buffer and pink indicates sample. The gradient formation and its relationship to analyte concentration at the flow cell is illustrated using five simulated snapshots (t_{start}–t_{stop}) of the injection line at different times (Courtesy: Sartorius)

The concept of SPR based screening platform has been part of drug discovery since early 2000 [3]. However, its importance in screening fragments was realized a decade ago [4]. The SPR based platform has often been challenged with fluorescent, luminescent, or radioactive based assays [5, 6]. Furthermore, these labeled assays failed to demonstrate any convincing beneficial effects to reduce the frequency of false hits arising out of HTS [7]. To overcome these vulnerabilities, much emphasis was made on the need for the development of a label-free screening methodology to evaluate fragments. The absence of reliable data on hit identification represents a drawback in differentiating between true hits from false positive hits. Also, most of the fragment libraries have autofluorescence fragments/compounds that makes these in vitro assay less reliable when compared to label-free platforms.

Fragment screening was typically hampered by a host of technical challenges including large numbers of samples, low molecular weight analytes (<250 Da), and weak affinity interactions (K_D: 10 μM to 10 mM) [8]. PioneerFE SPR OneStep Injections significantly improved SPR-based fragment screening, by increasing sample throughput through the generation of infinite gradients, while decreasing sample preparation time (Fig. 1.1). The technique was designed to streamline binding analysis by testing a multiple concentration series in a single injection. This not only saved sample preparation time but also reduced analyte consumption while minimizing human error by eliminating the preparation of multiple sample dilutions. Secondary screening can be completely avoided as fragment candidate selection is optimized during primary screening. Fragment screening with OneStep:

- Allow users to make decisions early on by obtaining reliable affinity (K_D) and kinetics (kon, koff) data directly from the primary screen.
- Provide fast time to first results, is fully automated and requires minimal assay development to arrive at the correct fragment candidates, and provides reproducible identification of fragment actives.

The data generated in a weak-affinity binding typically associated with fragments resembled a dose response plot and can be fitted with a real-time equilibrium binding model. The fragment off-rates are often fast enough to allow for a steady-state approximation. The Pioneer analysis software provides models incorporating kinetics, mass-transport corrections, and multi-site binding parameters that can adequately describe different interactions. Therefore, primary screening data was ready for K_D analysis without the need for laborious secondary screening and extra sample preparation steps.

There are several protocols and literature available for screening fragments from a collection of a library using SPR. We tried to provide detailed information from identification of fragments using virtual screen, production of tag specific proteins, ligation of proteins onto sensors, screening, and finally analysis to identify hits. Each chapter is subjected to protocol details, which are established at Jubilant Biosys. These protocols are ready to use and will be helpful for pharmaceutical, biotech, and university scientists involved in HTS screening using SPR.

References

1. Renaud J-P, Chung C-W, Helena Danielson U et al (2016) Biophysics in drug discovery: impact, challenges and opportunities. Nat Rev Drug Discov 15(10):679–698
2. SPR Instrument (2020). https://www.sprpages.nl/instruments
3. Thurmond RL, Wadsworth SA, Schafer PH et al (2001) Kinetics of small molecule inhibitor binding to p38 kinase. Eur J Biochem 268:5747–5754
4. Bergsdorf C, Ottl J (2010) Affinity-based screening techniques: their impact and benefit to increase the number of high quality leads. Expert Opin Drug Discov 5(11):1095–1107
5. Ediriweera MK, Tennekoon KH, Samarakoon SR (2019) In vitro assays and techniques utilized in anticancer drug discovery. J Appl Toxicol. 39(1):38–71
6. Gonzalez-Nicolini V, Fussenegger M (2005) In vitro assays for anticancer drug discovery–a novel approach based on engineered mammalian cell lines. Anticancer Drugs. 16(3):223–228
7. Sink R, Gobec S, Pečar S et al (2010) False positives in the early stages of drug discovery. Curr Med Chem 17(34):4231–4255
8. Ben J (2013) Davis, Daniel a. Erlanson (2013). Learning from our mistakes: the 'unknown knowns' in fragment screening. Bioorg Med Chem Lett 23(10):2844–2852. https://doi.org/10.1016/j.bmcl.2013.03.028

Principle of Surface Plasmon Resonance (OneStep)

2

Ramachandraiah Gosu and Sameer Mahmood Zaheer

Abstract

This chapter details the basic principle of SPR with appropriate components needed for the functioning of the system PioneerFE from Sartorius (previously Fortebio). SPR technology can be used to probe for affinity and kinetics between a range of molecules from small-molecule fragments to virus particles without the need for labels for detection. Binding interactions are monitored in a biosensor that is comprised of a thin gold-coated glass slide.

In most cases, a dextran matrix is attached to the gold coating, which acts as a substrate to which specific immobilization chemistries can be introduced to attach one binding partner to the biosensor. The other binding partner will be introduced to the biosensor using an automated fluidic system.

Keywords

Biosensor · Immobilization · Refractive index · Total internal reflection · Resonance

2.1 Principle

When light is made to pass through the biosensor surface and is reflected off the gold coating, at a certain angle of incidence, a portion of the light energy couples through the gold coating and creates a surface plasmon wave at the sample and the gold surface interface. The angle of incident light required to sustain the surface plasmon

R. Gosu (✉) · S. M. Zaheer
Structural Biology Division, Discovery Biology, Jubilant Biosys Ltd., Bengaluru, Karnataka, India
e-mail: ramachandraiah.gosu@jubilanttx.com; Sameer.Mahmood@jubilantbiosys.com

S. M. Zaheer, R. Gosu (eds.), *Methods for Fragments Screening Using Surface Plasmon Resonance*, https://doi.org/10.1007/978-981-16-1536-8_2

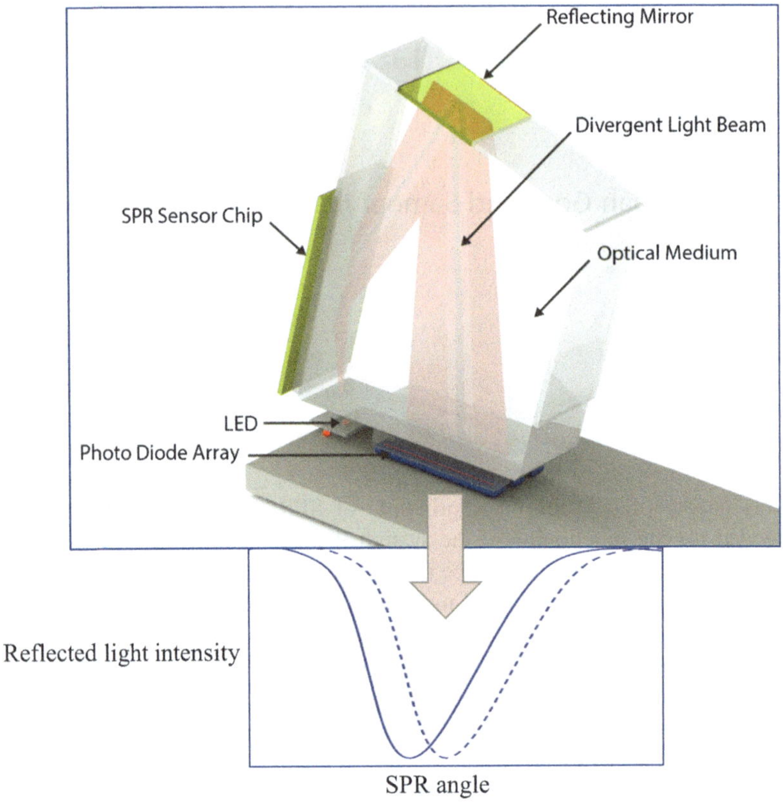

Fig. 2.1 SPR detection system—Kretschmann configuration (Courtesy: Sartorius)

wave is sensitive to refractive index changes that is proportional to mass changes at the surface. (Fig. 2.1).

The Pioneer instruments use SPR to measure in real-time molecular interactions as they occur at a gold sensor surface. This detection technology requires that one of the participants in the interaction is tethered to the surface and a sample, containing the second participant (occasionally multiple participants), is passed via flow across this surface.

SPR results from the excitation of polaritons by evanescent tunneling of light, of an appropriate wavelength and polarization, which is made incident to a metal (semitransparent noble metal film) coated dielectric substrate under conditions of total internal reflection (TIR). Changes in the position of the resonance with respect to incidence angle, or wavelength, are caused by proportional changes in the refractive index of a thin film on the metal surface, assuming the refractive index of the incident medium is constant. Biomolecules binding to the metal surface will cause such a refractive index change allowing the progress on the binding event to be monitored by tracking the SPR signal. (Fig. 2.2).

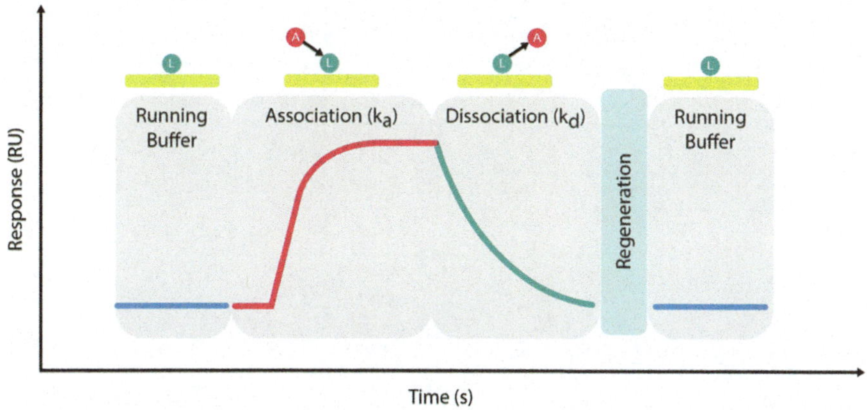

Fig. 2.2 Response curve showing course of biomolecular interaction. k_a (aka k_{on}) is the association rate constant and represents the association phase of the interaction (i.e. analyte injection). k_d (aka k_{off}) is the dissociation rate constant and represents the dissociation phase of the interaction (Courtesy: Sartorius)

In prism coupled SPR instruments, the dielectric substrate is a thin gold film coated onto a glass chip that is in optical contact with the TIR prism. The gold-coated side of the SPR chip is coated with surface chemistry and is in fluid contact with a nanoliter scale flow cell.

The evanescent wave that enables the energy transfer into the metal film exhibits exponential decay with distance from the interface and is therefore sensitive to a penetration depth of ~300 nm from the sensor surface. Gold is used in the sensor chips because it combines favorable SPR characteristics with chemical stability.

The sensor chip has two important functions:

1. It provides the physical conditions needed to produce the SPR signal.
2. It provides the biochemical conditions necessary for the interaction of interest to occur at the chip surface.

The specificity of SPR analysis is determined by the ligand's properties and by how it is attached to the sensor chip surface. Biomolecules may be attached to the surface via:

1. Covalent immobilization.
2. Affinity capture (the ligand is captured by another molecule that is bound (typically covalently) to the surface.
3. Hydrophobic adsorption.

Part I

SPR Compatible Protein Preparation

Preparation of Protein with Histidine Tag for Amine Coupling

3

Dinesh Thiagaraj and Ponni Vijayan

Abstract

Preparation of protein for SPR is a primary and important step to initiate the binding studies. For ease of purification of recombinant protein, fusing with a relevant tag is critical. These tags can be used for ligating the protein onto the sensor with specific sensor chemistry. This chapter will outline the preparation of histidine tagged protein and can be used for the amine coupling technique.

Keywords

IMAC · Ni-NTA · Histidine · Amine coupling · His tag

3.1 Introduction

One of the widely used label-free optical technologies is SPR, to study ligand (protein) and analyte (small molecule) binding kinetics in real-time in drug discovery research [1, 2] necessitates the availability of pure, homogenous, folded, and functionally active recombinant protein.

Poly-histidine tag is a small tag with high affinity to Ni-NTA resin, can be fused either at N-terminus or C-terminus depending on the protein to be studied and its folding, as the occlusion of the tag may be a hindrance for purification.

Purification of histidine tagged protein using the principle of chelation affinity, hence termed Immobilized metal affinity chromatography (IMAC), with Ni-NTA resin expedites the purification and confers better purity in the first step, and therefore, it is one of the widely used and suitable first steps in purification process [3, 4]. Although Ni-affinity itself is less specific compared to other widely used

D. Thiagaraj (✉) · P. Vijayan
Structural Biology Division, Discovery Biology, Jubilant Biosys Ltd., Bengaluru, Karnataka, India

S. M. Zaheer, R. Gosu (eds.), *Methods for Fragments Screening Using Surface Plasmon Resonance*, https://doi.org/10.1007/978-981-16-1536-8_3

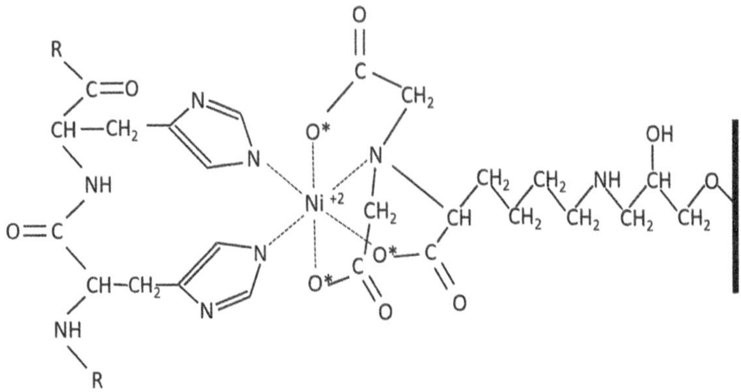

Fig. 3.1 Interaction model between poly-histidine tag and Ni-NTA (Source: Jubilant Biosys)

affinity techniques, Ni-affinity suits better than rest for large-scale purification processes [5, 6].

Of Ni-affinity chelation chemistries, Ni-NTA (Nickel—Nitrilotriacetic acid) is more specific compared to Ni-IDA (Nickel—Iminodiacetic acid) or Cu-IDA (Copper—Iminodiacetic acid) since NTA is a tetradentate ligand and Ni has six coordination sites that two sites would be free to bind histidine while copper has five coordination sites (Fig. 3.1). IDA is tridentate and TED is pentadentate, which leads to a higher rate of chelation and low affinity to histidine respectively with both Ni^{2+} and Cu^{2+} [7]. Hexa histidine is generally long enough for the tag to immobilize the protein to the resin or even to the sensor matrix if histidine capture strategy is used in SPR assay, however, 8x His or 10x His may be used in cases where purity needs further improvement, by increasing imidazole washes during Ni-affinity chromatography [8].

Optimizing lysis protocol is essential in the purification process (Table 3.1). While sonication being intense, micro fluidizing, homogenization qualitatively lyses better where sonication of prone to aggregate protein may cause soluble aggregation [9].

Both the highly pure, homogenous, properly folded protein and the suitable binding chemistry is essential to generate high-quality data. Partially denatured, soluble aggregated and misfolded protein lead to a higher rate of false positive hits during screening.

Amine coupling is one of the widely used coupling chemistries in SPR [10] and His tag protein can be coupled to sensor chip by histidine capture or the combination of both histidine capture and amine coupling as previously described [11].

This chapter presents procedures and methodologies involved in producing finer quality protein to be characterized in SPR assay.

Table 3.1 Troubleshooting for preparation of protein with Histidine Tag for amine coupling

Problem	Cause	Action
Soluble aggregation and low purity of protein after Ni-affinity step	• Intense sonication prone to aggregation nature of protein caused the soluble aggregation	• Optimization of sonication parameters (amplitude, on/off cycle, total time, temperature) • Detergent (triton-X-100) included in the resuspension buffer
Heterogeneity in protein population	• Soluble aggregates were not separated in Ni-affinity chromatography • Concentrating protein to inject in SEC caused the further aggregation • Resolution of separation of aggregates in HiLoad 16/600 Superdex 75 column was poor for this protein	• Optimization of Ni affinity chromatography and size exclusion chromatography where HiLoad 16/600 Superdex 200 was finally used to achieve a better resolution to separate soluble aggregates • Percentage of glycerol increased to 10% in an additional desalting step
Catalytic activity is less	• Higher ionic strength of NaCl and other salts	• Slowly dialyzed out the salt concentration
Protein aggregation during concentration	• Protein starts aggregating as it gets concentrated	• Reducing the RCF and run time helped with gentle mixing between runs

3.2 Materials

3.2.1 Cloning and Expression

1. Rosetta™ 2 DE3 pLysS Competent Cells (Cat. No.: 71403, Novagen).
2. *2xYT Media*
   (a) Bacto Tryptone (Cat. No.: 1612.00, Conda Labs).
   (b) Yeast Extract (Cat. No.: 1702.00, Conda Labs).
   (c) NaCl (Cat. No.: Q27608, Fisher Scientific).
3. 1000X Kanamycin (50 mg/mL) (Cat. No.: 0408-100G, VWR)
4. 1000X Chloramphenicol (33 mg/mL) (Cat. No.: C0378-100G, Sigma)
5. IPTG - Isopropyl β-D-1-thiogalactopyranoside (Stock – 1 M) (Cat. No.: 420322, CalBiochem).
6. Optima SP300 plus UV Spectrophotometer.
7. Beckman Coulter Avanti JXN-26 Centrifuge (Cat. No.: B34183).
8. Shell Lab Refrigerated Incubator shaker SI9R-2.

3.2.2 Purification of Proteins

1. Sonicator (Hielscher Ultrasonic processor, Cat. No.: UIP500hdT).
2. Beckman Coulter Avanti JXN-26 Centrifuge (Cat. No.: B34183).

3. Oakridge Tubes (Cat. No.: 541040, Tarsons).
4. Ni Sepharose 6 Fast Flow (Cat. No.: 17–5318-03, 500 mL).
5. XK 16/20 Column (Cat. No.: 28988937, GE).
6. XK 26/40 Column (Cat. No.: 28988949, GE).
7. Sephadex G-25 Medium (Cat. No.: 17003302, GE).
8. HiLoad 16/600 Superdex 200 prep grade (Cat. No.: 28989335, GE).
9. Vivaspin Turbo 15 PES, 10,000 MWCO PES (Cat. No.: VS15T01, Sartorius).
10. Whatman Grade 1 qualitative filter paper (Cat. No.: 1001–917, GE).
11. Dialysis tubing cellulose membrane (Cat. No.: D9652-100FT, Sigma, MW Cut-off: 14000 Da).
12. 20 mL of Ni Sepharose 6 Fast Flow resin packed in XK 16/20 column
13. 216 mL of Sephadex G-25 Medium resin packed in XK 26/40 column.
14. AKTA pure 25 (Cat. No.: 29–0182-28, GE).

3.2.3 Buffers Component Required for Protein Purification

1. HEPES (N-2-Hydroxyethylpiperazine-N-2-Ethane Sulfonic Acid) (Cat. No.: 11344041, Gibco).
2. NaCl (Cat. No.: Q27608, Fisher Scientific).
3. Tri-Sodium Citrate (Cat. No.: 14005, Fisher Scientific).
4. L-Arginine (Cat. No.: A0526, TCI Chemical).
5. L-Glutamic acid (Cat. No.: G1251-500G, Sigma).
6. Imidazole (Cat. No.: I202-500G, Sigma).
7. Glycerol (Cat. No.: Q15455, Fisher Scientific).
8. Triton-X-100 (Cat. No.: T8787, Sigma).
9. Protease Inhibitor Cocktail.
   (a) Aprotinin from bovine lung, Sigma, Cat. No.: A1153-100MG.
   (b) Leupeptin, Sigma, Cat. No.: L2884-100MG.
   (c) Pepstatin A, Sigma, Cat. No.: P5318-25MG).
10. Benzonase Nuclease (Cat. No.: E1014-25KU, Sigma).
11. KH_2PO_4 (Cat. No.: 13405, Fisher Scientific).
12. DL-Dithiothreitol (Cat. No.: D9779, Sigma).

3.2.4 2xYT Media Preparation

1. 16 g of Bacto Tryptone, 10 g of Yeast Extract, and 5 g of, NaCl were weighed and transferred to a 1000 mL beaker.
2. 1000 mL of MilliQ water was added and stirred well until a clear solution is obtained.
3. 500 mL of media was transferred into a 2000 mL conical Flask and Autoclaved at 121 °C, 15 psi pressure for 30 min.

3.2.5 Buffer Stocks Required for Protein Purification

1. *1 M HEPES*: 238.3 g of HEPES dissolved 800 mL milliQ water, pH adjusted to 8.0 with 10 N NaOH, and volume made up to 1000 mL with milliQ water.
2. *5 M NaCl*: 292.2 g NaCl dissolved in milliQ water to the final volume of 1000 mL.
3. *2 M Imidazole*: 136.1 g of Imidazole dissolved in milliQ water to the final volume of 1000 mL.
4. *Protease Inhibitor Stock (100 X)*: 50 mg Aprotinin from bovine lung, 68 mg Leupeptin dissolved in 97 mL of milliQ water, 34 mg Pepstatin A dissolved in 3 mL of methanol was added to make up to 100 mL solution.
5. *1 M IPTG*: 2.3 g of IPTG dissolved in milliQ water to the final volume of 10 mL
6. *1 M DTT*: 1.5 g of DTT dissolved in milliQ Water to the final volume of 10 mL.

3.2.6 Resuspension/Binding Buffer (1000 mL)

1. 50 mL of HEPES from 1 M Stock (final conc. 50 mM),
2. 100 mL of NaCl from 5 M Stock (final conc. 500 mM),
3. 29.4 g Tri-Sodium Citrate salt (final conc. 100 mM),
4. 5.2 g of L-Arginine (final conc. 30 mM),
5. 2.9 g of L-Glutamic acid (final conc. 20 mM),
6. 50 mL of Glycerol solution (5%),
7. pH adjusted to 7.5 and volume made up to 1000 mL with milliQ water.
8. Filtered through 0.8μm size filter.
9. 1X Protease Inhibitor cocktail, 1.5 U/mL Benzonase Nuclease, and 0.1% v/v Triton-X-100 was added to this resuspension buffer.

3.2.7 Ni-Affinity Wash Buffer (1000 mL)

1. 50 mL of HEPES from 1 M Stock (final conc. 50 mM)
2. 100 mL of NaCl from 5 M Stock (final conc. 500 mM)
3. 29.4 g Tri-Sodium Citrate salt (final conc. 100 mM)
4. 5.2 g of L-Arginine (final conc. 30 mM)
5. 2.9 g of L-Glutamic acid (final conc. 20 mM)
6. 50 mL of Glycerol solution (5%)
7. pH adjusted to 7.5 and made up to 1000 mL with milliQ water,
8. Filtered through 0.8 μm size filter.

3.2.8 Ni-Affinity Elution Buffer (1000 mL)

1. 50 mL of HEPES from 1 M Stock (final conc. 50 mM),
2. 100 mL of NaCl from 5 M Stock (final conc. 500 mM),

3. 29.4 g Tri-Sodium Citrate salt (final conc. 100 mM),
4. 5.2 g of L-Arginine (final conc. 30 mM),
5. 2.9 g of L-Glutamic acid (final conc. 20 mM),
6. 50 mL of Glycerol solution (5%),
7. 250 mL of Imidazole from 2 M Stock Solution (final conc. 500 mM).
8. pH adjusted to 7.5 and volume made up to 1000 mL with milliQ water,
9. Filtered through 0.8 μm size filter.

3.2.9 Desalting/SEC Buffer (1000 mL)

1. 50 mL of HEPES from 1 M Stock (final conc. 50 mM),
2. 100 mL of NaCl from 5 M Stock (final conc. 500 mM),
3. 14.7 g Tri-Sodium Citrate salt (final conc. 50 mM),
4. 8.7 g of L-Arginine (final conc. 50 mM),
5. 2.9 g of L-Glutamic acid (final conc. 20 mM),
6. 6.8 g of KH_2PO_4 (final conc. 50 mM)
7. 100 mL of Glycerol solution (10%),
8. pH adjusted to 7.5 and volume made up to 1000 mL with milliQ water,
9. Filtered through 0.8 μm size filter.

3.2.10 Dialysis Buffer (1000 mL)

1. 25 mL of HEPES from 1 M Stock (final conc. 25 mM),
2. 30 mL of NaCl from 5 M Stock (final conc. 150 mM)
3. 1 mL of DTT from 1 M Stock (final conc. 1 mM)
4. 0.174 g of L-Arginine (final conc. 1 mM)
5. 100 mL of Glycerol (10%)
6. pH adjusted to 7.5 and volume made up to 1000 mL with milliQ water,
7. Filtered through 0.8 μm size filter.

3.3 Methods

3.3.1 Gene Synthesis and Cloning

Gene constructs with desired boundaries designed with 6× His tag at N-terminus. Gene constructs synthesized and cloned (see **Note 1**).

3.3.2 Transformation and Expression

1. 100 ng of plasmid (gene of interest in pET28a vector) was added to 50 μL of Rosetta™ 2 DE3 pLysS competent cells and incubated for 30 min in ice.

2. The cells were heat shocked at 42 °C with preheated water bath for 75 s.
3. The cells were kept in ice for 5 min.
4. 50 μL of competent cells were transferred to 1 mL of 2 × YT media and incubated for 1 hour at 37 °C.
5. Post incubation, the samples were centrifuged at 5000 g for 10 min at 24 °C.
6. Supernatant was discarded and the pellet resuspended in 100μL of 2 × YT media.
7. 100 μL of transformed cells were plated onto LB agar plate with 1X kanamycin and 1X chloramphenicol and incubated for 16 hours at 37 °C
8. Single colony after 16 hours incubation was inoculated in 10 mL 2 × YT media, subsequently reinoculated in 100 mL (primary culture) with 1× kanamycin and 1× chloramphenicol.
9. 10 flasks of 500 mL culture media in 2000 mL borosilicate conical flask were inoculated with 2% primary culture which had an OD of 1.9 measured at 600 nm using a spectrophotometer.
10. Inoculated secondary culture was further incubated at 37 °C at 200 rpm in a shaker incubator.
11. Wait till the culture reaches 0.6 OD measured at 600 nm, then add 0.5 mL of IPTG from 1 M Stock to these flasks.
12. Post induction, the flasks were incubated for 16 h at 18 °C in a shaker incubator adjusted to 180 rpm.
13. After 16 hours, the cells were harvested by centrifugation at 7000 g for 10 mins at 4 °C.
14. The harvested cell pellet (~40 g) stored at −80 °C.

3.3.3 Purification Process

3.3.3.1 Lysate Preparation

1. Stored cell pellet was again suspended in resuspension buffer with a ratio of 1:10 (weight/volume) – cell pellet versus resuspension buffer (i.e., 40 g of cell pellet suspended in 400 mL of resuspension buffer). (see **Note 2**).
2. The cell suspension was placed in 4 °C, sonicated with 40% amplitude, for 10 s on/5 s off, for 20 min at 200 rpm with stirring. (see **Note 3**).
3. Lysate was clarified by centrifugation at 39,000 g for 1 h in Oak Ridge tubes in Beckman Coulter Avanti JXN-26 centrifuge. (see **Note 4**).
4. Supernatant was decanted and filtered using Whatman Filter paper No 1. (see **Note 5**).

3.3.3.2 Column Packing

20 mL of Ni-Sepharose FF resin was packed in XK 16/20 column, washed with 200 mL milliQ water, and equilibrated with 100 mL resuspension buffer.

3.3.3.3 Akta Pure FPLC System Wash

Inlet A3, A1 and B1 were washed with milliQ water and equilibrated with resuspension buffer, wash buffer, and elution buffer sequentially.

3.3.3.4 Chromatography

1. 380 mL of sample was loaded onto Ni affinity resin column with a flow rate of 1 mL/min through inlet A3. (see **Note 6**)
2. Lysate collected via Outlet 1.
3. 0–0 mM Imidazole wash—200 mL (1 mL/min flow rate) through inlet A1, flow through collected in fractions (9 mL)
4. 0–30 mM gradient wash—200 mL (1 mL/min flow rate) through inlet A1 and B1, flow through collected in fractions (9 mL)
5. 30–500 mM gradient elution—200 mL (1 mL/min flow rate) through inlet A1 and B1, flow through collected in fractions (5 mL). (Fig. 3.2)
6. Eluted Fractions were evaluated in SDS-PAGE and confirmed in Western blot (Fig. 3.3), pooled, and desalted (for imidazole) in 216 mL Desalting column (XK 26/40. The desalted protein sample concentrated to 5 mL and injected into HiLoad 16/600 Superdex 200 prep grade column. (see **Note 7**).
7. The elution fractions evaluated in SDS-PAGE were pooled, and dialyzed against dialysis buffer with 1000 mL batch at 4 °C with 2 buffer exchanges in first 2 hours and extended for 16 hours in total.

3.3.3.5 Dialysis

1. Required length of the dialysis bag was heated in milliQ water with a pinch of sodium bicarbonate at 90 °C for 5 min.
2. Once the water was cooled down, the dialysis bag was rinsed with water to remove residual sodium bicarbonate.
3. Protein was loaded onto dialysis bag, the ends were clamped and stringed up in a glass beaker containing 1000 mL of dialysis buffer. (see **Note 8**).
4. The dialysed protein was concentrated to 4 mg/mL using Vivaspin Turbo15 10 kDa concentrators at 4500 g in Eppendorf 5810R centrifuged at 4 °C, aliquoted and frozen in −80 °C until use. (see **Note 9**).

3.4 Notes

1. Gene synthesis is outsourced – Hence will not be discussed here.
2. Efficiency of cell lysis was more when frozen cell pellets were used.
3. Temperature rise during the sonication process should be avoided.
4. Before spinning the sample, make sure the rotor temperature is maintained at 4 °C.
5. Inappropriate filter leads to column clogging while processing.
6. Pressure should not exceed 0.4 MPa or as per manufacturer's instruction for pre-packed column.
7. Protein eluted at 0.2 mL/min flow rate, with 2 mL fraction collection.

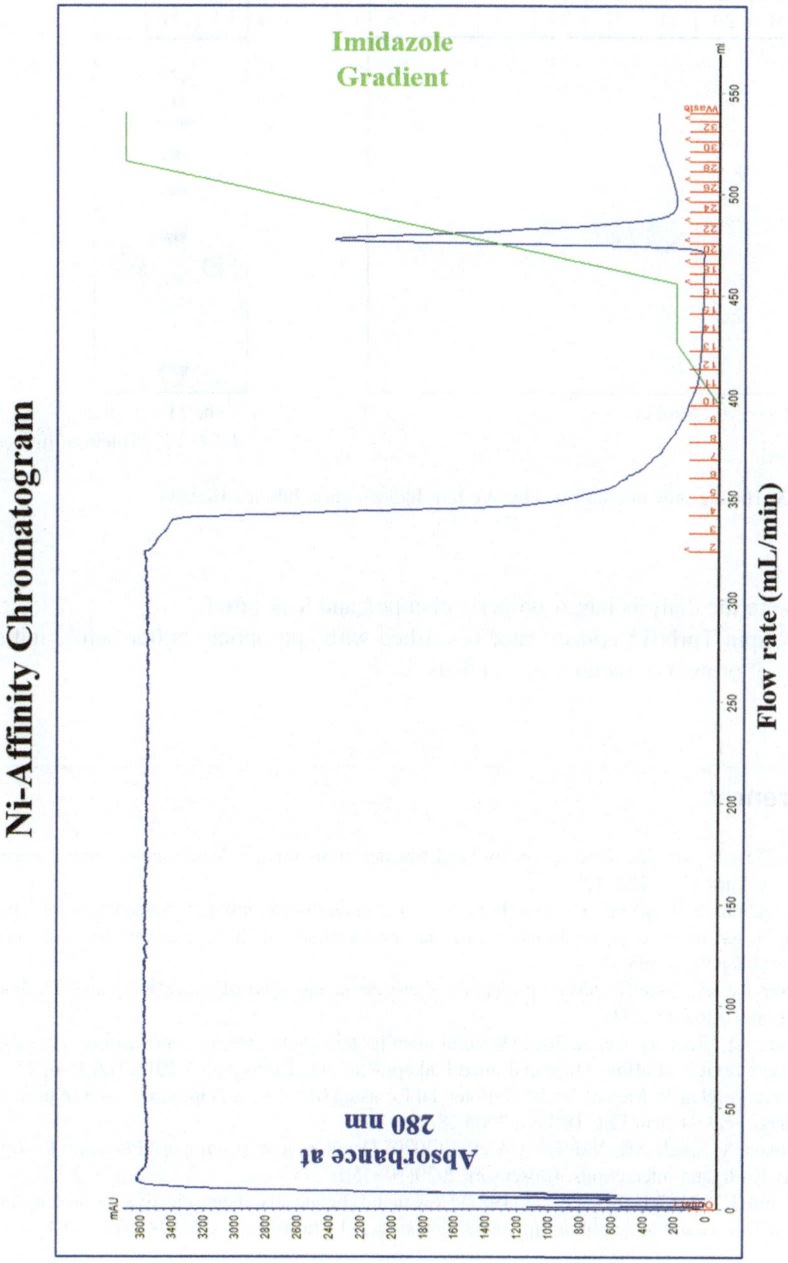

Fig. 3.2 Ni-Affinity Chromatography of protein of interest (Source: Jubilant Biosys)

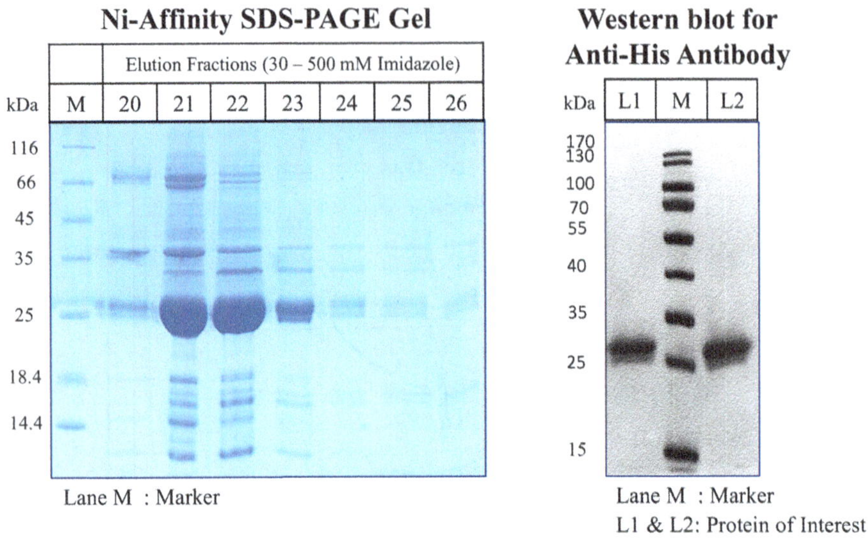

Fig. 3.3 His tag protein confirmed by Western blot (Source: Jubilant Biosys)

8. Ensure the dialysis bag is properly clamped and leak-proof.
9. Vivaspin Turbo15 concentrator is washed with appropriate buffer before initiation of protein concentration via dialysis.

References

1. Sun Y-S (2013) Optical biosensors for label-free detection of biomolecular interactions. Instrum Sci Technol 42(2):109–127
2. Vaisocherová H, Brynda E, Homola J (2015) Functionalizable low-fouling coatings for label-free biosensing in complex biological media: advances and applications. Anal Bioanal Chem 2015(407):3927–3953
3. Bornhorst JA, Falke JJ (2000) Purification of proteins using polyhistidine affinity tags. Methods Enzymol 326:245–254
4. Young CL, Zachary T et al (2012) Recombinant protein expression and purification: a comprehensive review of affinity tags and microbial applications. Biotechnol J 2012(7):620–634
5. Garerc-Porekar V, Menart V (2005) Potential for using histidine tags in purification of proteins at large scale. Chem Eng Technol 2005:28. No. 11
6. Boonen A, Singh AK, Van Hout A et al (2020) Development of a novel SPR assay to study CXCR4-ligand interactions. Biosensors 2020(10):150
7. Hochuli E, Döbeli H, Schacher A (1987) New metal chelate adsorbent selective for proteins and peptides containing neighbouring histidine residues. J Chromatogr 411(1987):177–184

8. Döbeli H (2020) Polyhistidine affinity chromatography: from the idea to a method of broad applicability and the pitfalls in-between. Chimia 2020:74. No. 5
9. Kwon SB, Yu JE, Kin J et al (2019) Quality screening of incorrectly folded soluble aggregates from functional recombinant proteins. Int J Mol Sci 2019(20):907
10. Fischer MJE (2010) In: Mol NJ, Fischer MJE (eds) Methods and protocols. Humana Press, Totowa, NJ, pp 55–73
11. Wang X, Liu Q, Tan X et al (2019) Covalent affixation of histidine-tagged proteins tethered onto Ni-nitrilotriacetic acid sensors for enhanced surface plasmon resonance detection of small molecule drugs and kinetic studies of antibody/antigen interactions. Analyst 144(2):587–593

Preparation of Protein with AviTag™ for Biotin-Based Capture

4

Saravanan Kandan, Dinesh Thiagaraj, and Rashmi Rekha Devi

Abstract

Avidin tagged proteins take advantage where amine coupling fails to provide required interactions between protein and analyte. Immobilization of biomolecules by avidin–biotin-based methods is very common due to the robustness of the interaction. Preparation of AviTag™ protein for SPR is the primary work to initiate the binding studies. This chapter will outline the preparation of AviTag™ protein.

Keywords

Biotin · Avidin · AviTag · Biotinylation · BirA biotin ligase

4.1 Introduction

While the amine coupling is advantageous due to irreversibility, it requires higher protein concentration and the coupling efficiency with EDC/NHS is low [1]. Multiple amines in the same protein are involved in amine coupling leading to different orientations of immobilization of protein in the sensor chip [2, 3]. Moreover, when the lysine is involved in active site catalysis or binding pocket, amine coupling is incompatible. In these cases, an alternative biotin capture methodology can be used that harnesses the strong natural non-covalent interaction between avidin and biotin with an equilibrium dissociation constant K_D of 10^{-12} M to 10^{-14} M [4] to immobilize ligand to the sensor chip. One of the strategies to carry out this coupling method involves producing biotin labeled protein.

S. Kandan (✉) · D. Thiagaraj · R. R. Devi
Structural Biology Division, Discovery Biology, Jubilant Biosys Ltd., Bengaluru, Karnataka, India
e-mail: saravanan.kandan@jubilantbiosys.com

S. M. Zaheer, R. Gosu (eds.), *Methods for Fragments Screening Using Surface Plasmon Resonance*, https://doi.org/10.1007/978-981-16-1536-8_4

Table 4.1 Troubleshooting for preparation of protein with avidin tag for biocap capture

Problem	Cause	Action
Lower percentage of biotinylation	• Overexpression of desired protein relative to BirA enzyme	• Optimization of biotin, IPTG, incubation temperature, RPM performed
Lower percentage of biotinylation	• Occlusion of AviTag™ when placed in N-terminal (after histidine tag)	• Designed construct with AviTag™ in C-terminal

Biotinylation can be done with *E. coli* BirA biotin ligase enzyme. AviTag™ is a 15 amino acid (GLNDIFEAQKIEWHE) acceptor peptide with a lysine residue and is a nonnatural substrate for BirA enzyme [5]. BirA catalyzes the ligation of biotin to the amine group of lysine in AviTag™ with ATP as cofactor. This chapter presents the procedure for in vivo biotinylation of desired protein by co-expressing BirA biotin ligase in *E.coli* (Table 4.1). The biotinylated protein was immobilized on avidin coated sensor chip via biotin–avidin interaction.

4.2 Materials

4.2.1 Cloning and Expression

1. BL21 Competent cells with IPTG-inducible BirA Expression plasmid (Cat. No.: 27462, BPS Biosciences).
2. *2xYT Media*
   (a) Bacto Tryptone (Cat. No.: 1612.00, Conda Labs).
   (b) Yeast Extract (Cat. No.: 1702.00, Conda Labs).
   (c) NaCl (Cat. No.: Q27608, Fisher Scientific).
3. 1000× Kanamycin (50 mg/mL) (Cat. No.: 0408-100G, VWR)
4. 1000× Streptomycin (100 mg/mL) (Cat. No.: S6501-100G, Sigma)
5. 100× Biotin (5 mM, 1.221 mg/mL) (Cat. No.: B4639-1G, Sigma),
6. IPTG—Isopropyl β-D-1-thiogalactopyranoside (Stock – 1 M) (Cat. No.: 420322, CalBiochem).
7. Optima SP300 plus UV Spectrophotometer.
8. Beckman Coulter Avanti JXN-26 Centrifuge (Cat. No.: B34183).
9. Shell Lab Refrigerated Incubator shaker SI9R-2.

4.2.2 Purification of Proteins

1. Sonicator (Hielscher Ultrasonic processor, Cat. No.: UIP500hdt).
2. Beckman Coulter Avanti JXN-26 Centrifuge (Cat. No.: B34183).
3. Oakridge Tubes (Cat. No.: 541040, Tarsons).
4. Ni Sepharose 6 Fast Flow (Cat. No.: 17–5318-03, 500 mL).
5. XK 16/20 Column (Cat. No.: 28988937, GE).

6. XK 26/40 Column (Cat. No.: 28988949, GE).
7. Sephadex G-25 Medium (Cat. No.: 17003302, GE).
8. HiLoad 16/600 Superdex 200 pg (Cat. No.: 28989335, GE).
9. Vivaspin Turbo 15 PES, 10,000 MWCO PES (Cat. No.: VS15T01, Sartorius).
10. Whatman Grade 1 qualitative filter paper (Cat. No.: 1001–917, GE).
11. Dialysis tubing cellulose membrane (Cat. No.: D9652-100FT, Sigma, MW cutoff: 14,000 Da).
12. 20 mL of Ni Sepharose 6 Fast Flow resin packed in XK 16/20 column
13. 216 mL of Sephadex G-25 Medium resin packed in XK 26/40 column.
14. AKTA pure 25 (Cat. No.: 29–0182-28, GE).

4.2.3 Buffers Component Required for Protein Purification

1. HEPES (N-2-Hydroxyethylpiperazine-N-2-Ethane Sulfonic Acid) (Cat. No.: 11344041, Gibco).
2. NaCl (Cat. No.: Q27608, Fisher Scientific).
3. Tri-Sodium Citrate (Cat. No.: 14005, Fisher Scientific).
4. L-Arginine (Cat. No.: A0526, TCI Chemical).
5. L-Glutamic acid (Cat. No.: G1251-500G, Sigma).
6. Imidazole (Cat. No.: I202-500G).
7. Glycerol (Cat. No.: Q15455, Fisher Scientific).
8. Triton-X-100 (Cat. No.: T8787, Sigma).
9. Protease Inhibitor Cocktail (Aprotinin from bovine lung, Sigma, Cat. No.: A1153-100MG; Leupeptin, Sigma, Cat. No.: L2884-100MG; Pepstatin A, Sigma, Cat. No.: P5318-25MG).
10. Benzonase Nuclease (Cat. No.: E1014-25KU, Sigma).
11. KH_2PO_4 (Cat. No.: 13405, Fisher Scientific).
12. DL-Dithiothreitol (Cat. No.: D9779, Sigma).

4.2.4 2xYT Media Preparation

1. Weigh 16 g of Bacto Tryptone, 10 g of Yeast Extract, and 5 g of NaCl.
2. Add 1000 mL of milliQ water. Stir it until a clear solution is obtained and autoclave at 121 °C, 15 psi pressure for about 30 min.

4.2.5 Buffer Stocks Required for Protein Expression and Purification

1. *1 M HEPES*: 238.3 g of HEPES dissolved in 1000 mL of milliQ water, pH adjusted to 8.0 with 10 N NaOH.
2. *5 M NaCl*: 292.2 g NaCl dissolved in milliQ water to the final volume of 1000 mL.

3. *2 M Imidazole*: 136.1 g of Imidazole dissolved in milliQ water to the final volume of 1000 mL.
4. *Protease Inhibitor Stock (100×)*: 50 mg Aprotinin from bovine lung, 68 mg Leupeptin dissolved in 97 mL of milliQ water, 34 mg Pepstatin A dissolved in 3 mL of methanol was added to make up to 100 mL solution.
5. *1 M IPTG*: 2.381 g of IPTG dissolved in milliQ water to the final volume of 10 mL.
6. *1 M DTT*: 1.542 g of DTT dissolved in milliQ water to the final volume of 10 mL.

4.2.6 Preparation of Biotin (5 mM)

1. 5 mL from 1 M HEPES stock was added to 95 mL of milliQ water (final conc. 50 mM)
2. 122.16 mg of biotin was added to 100 mL of 50 mM HEPES solution, pH adjusted to 7.0
3. Biotin stock is filtered with 0.22μm size filter.

4.2.7 Resuspension/Binding Buffer (1000 mL)

1. 50 mL of HEPES from 1 M Stock (final conc. 50 mM)
2. 100 mL of NaCl from 5 M Stock (final conc. 500 mM)
3. 29.41 g Tri-Sodium Citrate salt (final conc. 100 mM)
4. 5.226 g of L-Arginine (final conc. 30 mM)
5. 2.942 g of L-Glutamic acid (final conc. 20 mM)
6. 50 mL of Glycerol solution (5%)
7. pH adjusted to 7.5 and volume made up to 1000 mL with milliQ water,
8. Filtered through 0.8μm size filter.
9. 1× Protease Inhibitor cocktail,
10. 1.5 U/mL Benzonase Nuclease,
11. 0.1% v/v Triton-X-100.

4.2.8 Ni-Affinity Wash Buffer (1000 mL)

1. 50 mL of HEPES from 1 M Stock (final conc. 50 mM),
2. 100 mL of NaCl from 5 M Stock (final conc. 500 mM),
3. 29.41 g Tri-Sodium Citrate salt (final conc. 100 mM),
4. 5.226 g of L-Arginine (final conc. 30 mM),
5. 2.942 g of L-Glutamic acid (final conc. 20 mM),
6. 50 mL of Glycerol solution (5%),
7. pH adjusted to 7.5,
8. Volume made up to 1000 mL with milliQ water.
9. Filtered through 0.8μm size filter.

4.2.9 Ni-Affinity Elution Buffer (1000 mL)

1. 50 mL of HEPES from 1 M Stock (final conc. 50 mM),
2. 100 mL of NaCl from 5 M Stock (final conc. 500 mM),
3. 29.41 g Tri-Sodium Citrate salt (final conc. 100 mM),
4. 5.226 g of L-Arginine (final conc. 30 mM),
5. 2.942 g of L-Glutamic acid (final conc. 20 mM),
6. 50 mL of Glycerol solution (5%),
7. 250 mL of Imidazole from 2 M Stock Solution (final conc. 500 mM).
8. pH adjusted to 7.5,
9. Volume made up to 1000 mL with milliQ water.
10. Filtered through 0.8μm size filter.

4.2.10 Desalting/SEC Buffer (1000 mL)

1. 50 mL of HEPES from 1 M Stock (final conc. 50 mM),
2. 100 mL of NaCl from 5 M Stock (final conc. 500 mM),
3. 14.7 g Tri-Sodium Citrate salt (final conc. 50 mM),
4. 8.71 g of L-Arginine (final conc. 50 mM),
5. 2.942 g of L-Glutamic acid (final conc. 20 mM),
6. 6.8 g of KH_2PO_4 (final conc. 50 mM)
7. 100 mL of Glycerol solution (10%),
8. pH adjusted to 7.5,
9. Made up to 1000 mL with milli Q water.
10. Filtered through 0.8μm size filter.

4.2.11 Dialysis Buffer (1000 mL)

1. 25 mL of HEPES from 1 M Stock (final conc. 25 mM),
2. 30 mL of NaCl from 5 M Stock (final conc. 150 mM)
3. 1 mL of DTT from 1 M Stock (final conc. 1 mM)
4. 0.174 g of L-Arginine (final conc. 1 mM)
5. 100 mL of Glycerol (10%)
6. pH adjusted to 7.5,
7. Made up to 1000 mL with milliQ water.
8. Filtered through 0.8μm size filter.

4.3 Methods

4.3.1 Cloning and Expression

1. Gene constructs with desired boundaries designed with.

(a) 6× His tag at N-terminus followed by AviTag™
(b) 6× His tag at N-terminus and AviTag™ at C-Terminus.

2. Gene constructs synthesized and cloned. (see **Note 1**).
3. Competent cells with 100 ng of gene construct in pET28a vector.
4. Incubated in ice for 30 min.
5. The cells were heat shocked at 42 °C with preheated water bath for 75 s.
6. The cells were kept in ice for 5 min.
7. 50μL of competent cells were transferred to 1 mL of 2 × YT media and incubated for 1 h at 37 °C.
8. Post incubation, the samples were centrifuged at 5000 g for 10 min at 24 °C.
9. Supernatant was discarded and pellet resuspended in 100 uL of 2 × YT media.
10. 100μL of transformed cells were plated onto LB agar plate with 1× kanamycin and 1× streptomycin.
11. Single colony inoculated in 10 mL 2 × YT media, subsequently reinoculated in 100 mL (Primary Culture) with 1× kanamycin and 1× Streptomycin.
12. 10 flasks of 500 mL culture in 2000 mL borosilicate conical flask were inoculated with 3% primary culture, which had an OD of ~1.8 measured at 600 nm using a spectrophotometer.
13. Inoculated secondary culture was further incubated at 37 °C at 200 rpm in a shaker incubator.
14. Wait till the culture reaches 0.6 OD measured at 600 nm, then add 0.5 mL of IPTG from 1 M Stock and 5 mL of 100× Biotin Stock to each flask.
15. Post induction, the flasks were incubated for 16 h at 18 °C in a shaker incubator adjusted to 180 rpm.
16. After 16 h, the cells were harvested by centrifugation at 7000 g for 10 min at 4 °C.
17. The harvested cell pellet (~30 g) stored at −80 °C.

4.3.2 Purification Process

4.3.2.1 Lysate Preparation

1. Stored cell pellet is suspended in resuspension buffer in 1:10 weight/volume ratio of cell pellet versus resuspension buffer (i.e., 30 g of pellet suspended in 300 mL of resuspension buffer). (see **Note 2**).
2. The cell suspension placed in 4 °C was sonicated with 40% amplitude, for 10 s on/5 s off, for 20 min at 200 rpm stirring (see **Note 3**).

3. Lysate was clarified by centrifugation at 39000 g for 1 h in Oak Ridge tubes in Beckman Coulter Avanti JXN-26 centrifuge. (see **Note 4**).
4. Supernatant was decanted and filtered using Whatman Filter paper No. 1 (see **Note 5**).

4.3.2.2 Column Packing

20 mL Ni-Sepharose FF resin was packed in XK 16/20 column, washed with 200 mL Milli Q water, and equilibrated with 100 mL resuspension buffer.

4.3.2.3 Akta Pure FPLC System Wash

Inlet A3, A1, and B1 were washed with MilliQ water and equilibrated with resuspension buffer, wash buffer, and elution buffer sequentially.

4.3.2.4 Chromatography

1. 280 mL of sample was loaded onto Ni affinity resin column with a flow rate of 1 mL/min through inlet A3. (see **Note 6**)
2. Lysate collected via Outlet 1.
3. 0–0 mM Imidazole wash—200 mL (1 mL/min flow rate) through inlet A1, flow through collected in fractions (9 mL)
4. 0–30 mM gradient wash—200 mL (1 mL/min flow rate) through inlet A1 and B1, flow through collected in fractions (9 mL)
5. 30–500 mM gradient Elution—200 mL (1 mL/min flow rate) through inlet A1 and B1, flow through collected in fractions (5 mL) (Fig. 4.1).
6. Eluted Fractions evaluated in SDS-PAGE and confirmed with Western blot (Fig. 4.2), pooled, and desalted (for imidazole) in 220 mL Desalting column (XK 26/40. The desalted protein sample concentrated to 5 mL and injected into HiLoad 16/600 Superdex 200 prep grade column. (see **Note 7**).
7. The elution fractions were evaluated in SDS-PAGE, pooled, and dialyzed against dialysis buffer with 1000 mL batch at 4 °C with 2 buffer exchanges in first 2 hours and extended for 16 hours in total.

4.3.2.5 Dialysis

1. Required length of the dialysis bag was heated in milliQ water with a pinch of sodium bicarbonate at 90 °C for 5 min.
2. Once the water was cooled down, the dialysis bag was rinsed with water to remove residual sodium bicarbonate.
3. Protein was loaded onto dialysis bag, the ends were clamped and stringed up in a glass beaker containing 1000 mL of dialysis buffer. (see **Note 8**).
4. The dialyzed protein was concentrated to 4 mg/mL using Vivaspin Turbo 15 10 kDa concentrators at 4500 g in Eppendorf 5810R centrifuged at 4 °C, aliquoted and frozen in −80 °C until use.

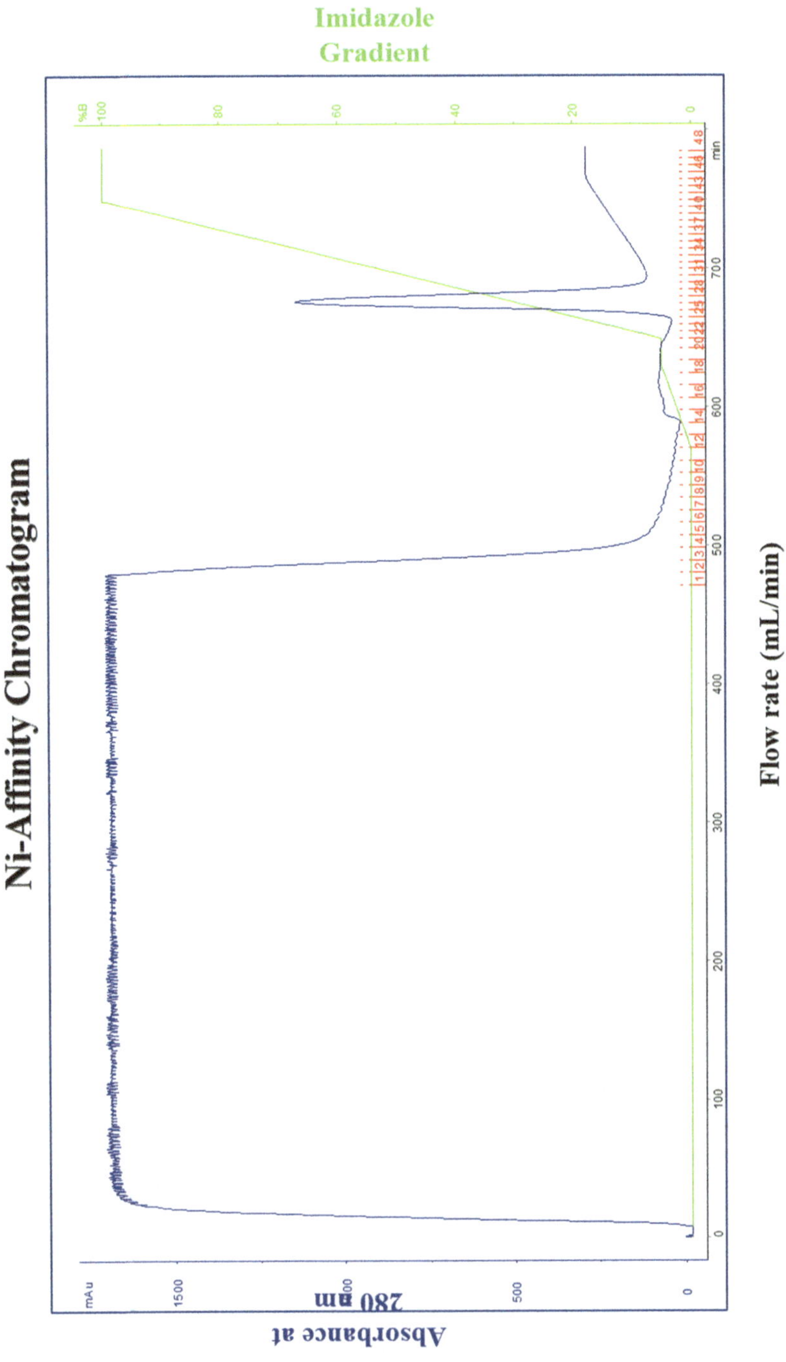

Fig. 4.1 Ni-affinity chromatography of biotinylated protein (Source: Jubilant Biosys)

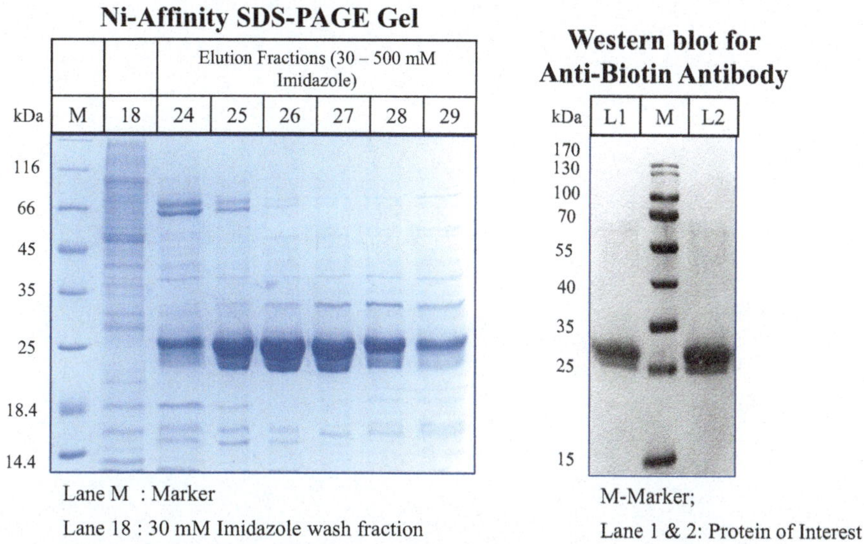

Fig. 4.2 Purified protein biotinylation confirmed with Western blot (Source: Jubilant Biosys)

4.4 Notes

1. Gene synthesis is outsourced—hence will not be discussed here.
2. Efficiency of cell lysis was more when frozen cell pellets were used.
3. Temperature rise during the sonication process should be avoided.
4. Before spinning the sample, make sure the rotor temperature is maintained at 4 °C.
5. Inappropriate filter leads to column clogging while processing.
6. Pressure should not exceed 0.4 MPa or as per manufacturer's instruction for pre-packed column.
7. Protein eluted at 0.2 mL/min flow rate, with 2 mL fraction collection.
8. Ensure the dialysis bag is properly clamped and leakproof.

References

1. Wang X, Zhou F (2018) Dual-valve and counter-flow surface Plasmon resonance. Anal Chem 90 (8):4972–4977
2. Johnson DL, Martin LL (2005) Controlling protein orientation at interfaces using histidine tags: an alternative to Ni/NTA. J Am Chem Soc 127(7):2018–2019
3. Karyakin AA, Presnova GV, Rubtsova MY et al (2000) Oriented immobilization of antibodies onto the gold surfaces via their native thiol groups. Anal Chem 72(16):3805–3811
4. Michael Green N (1990) (1990) Avidin and streptavidin. Methods Enzymol 184:51–67
5. Beckett D, Kovaleva E, Schatz PJ (1999) A minimal peptide substrate in biotin holoenzyme synthetase-catalyzed biotinylation. Protein Sci 8:921–929

Part II

Characterization of Protein for SPR

5 Protein Stability Using Thermal Shift Assay (TSA): pH Tolerance

Srinivasan Swaminathan and A. L. Theerthan

Abstract

SPR assay requires ligation of protein to biosensor chips by amine coupling method or biotin capture methods. However, prior to ligation of protein onto the sensors, protein of interest need to be evaluated for pH tolerability and stability. Buffer mismatch is an important aspect of SPR assay due to its high sensitivity for small changes. Protein storage buffer is relatively different from immobilization and running buffer. To perform amine coupling, protein needs to be ligated at acidic pH (<7) conditions. Thermal shift assay was used as a tool to evaluate the stability of proteins in these buffer conditions based on the Melt curve and Tm recorded in comparison to the protein storage buffer.

Keywords

Thermal Shift Assay (TSA) · Melting temperature (Tm) · Differential Scanning Fluorimetry (DSF) · Fluorescence

5.1 Introduction

Differential Scanning Fluorimetry (DSF) commonly referred to as Thermal Shift Assay (TSA) [1] is the most commonly used method to make a quick evaluation of protein stability based on the temperature at which protein melts in presence of fluorescent dye, Sypro orange. The principle of TSA is with increasing temperature, proteins unfold, leading to exposure of hydrophobic amino acids present in proteins. The Sypro orange dye binds to these hydrophobic amino acid residues of protein and

S. Swaminathan (✉) · A. L. Theerthan
Structural Biology Division, Discovery Biology, Jubilant Biosys Ltd., Bengaluru, Karnataka, India
e-mail: Srinivasan.S@jubilantbiosys.com

S. M. Zaheer, R. Gosu (eds.), *Methods for Fragments Screening Using Surface Plasmon Resonance*, https://doi.org/10.1007/978-981-16-1536-8_5

fluorescence is measured (Ex 492 nm/Em 610 nm) [2]. Increase in fluorescence is the direct measure of protein unfolding. Melting Temperature (Tm) corresponds to the inflection point of the melt curve where an equal population of folded and unfolded protein is inferred [2]. The experiment involves measuring change in Tm of the protein under varying buffering conditions in comparison with protein storage buffer. Each protein is tested at different pH conditions before starting an SPR assay. This chapter focuses on the development of TSA protocol to evaluate the optimal pH in buffer required for immobilization of protein.

5.2 Materials

5.2.1 Components Required for TSA

1. Sodium acetate trihydrate (Cat. No: S7670, Sigma-Aldrich).
2. SYPRO™ Orange Protein Gel Stain (5000× Concentrate in DMSO) (Cat. No: S6650, ThermoFisher Scientific).
3. Multiplate™ 96-Well PCR Plates, low profile, unskirted, clear - MLL9601 (Cat. No: MLL9601, BIO-RAD).
4. Microseal "B" PCR Plate Sealing Film, adhesive, optical (Cat. No: MSB1001, BIO-RAD).
5. Single channel pipettes—p2.5, p10, p20, p100 (Eppendorf).
6. CFX96™ Real-Time PCR Detection system—C1000 Thermal Cycler (Bio-Rad).
7. HEPES (Cat. No: H3375, Sigma-Aldrich).
8. Sodium chloride (Cat. No: S9888, Sigma-Aldrich).
9. Glycerol (99.5%) (Cat. No: G7893, Sigma-Aldrich).
10. TCEP hydrochloride (Cat. No: HR2–801, Hampton Research).
11. Ice bucket and ice.

5.2.2 Buffer and Reagents for TSA

1. *12 mM Sodium acetate (pH 4.5):* 16.32 mg Sodium acetate trihydrate powder made up to 10 mL with 4.5 pH titration using Conc. HCl/8 M NaOH
2. *12 mM Sodium acetate (pH 5.0):* 16.32 mg Sodium acetate trihydrate powder made up to 10 mL with 5.0 pH titration done using Conc. HCl/8 M NaOH
3. *12 mM Sodium acetate (pH 5.5):* 16.32 mg Sodium acetate trihydrate powder made up to 10 mL with 5.5 pH titration done using Conc. HCl/8 M NaOH
4. *12 mM Sodium acetate pH 6.0:* 16.32 mg Sodium acetate trihydrate powder made up to 10 mL with 6.0 pH titration done using Conc. HCl/8 M NaOH.
5. *Protein working stock* (50μM): 6.75μL of stock protein (222.4μM) diluted with the addition of 23.25μL of protein storage buffer to get 30μL of 50μM working stock protein solution.

6. *Protein storage buffer:* 50 mM HEPES pH 7.0, 500 mM NaCl, 5% Glycerol, 0.5 mM TCEP.
7. *1 M HEPES:* 2.38 g of HEPES powder weighed and made up to 10 mL solution with pH 7.0 (pH titration using Conc. HCl/8 M NaOH)
8. *5 M NaCl:* 2.92 g of NaCl weighed and made up to 10 mL solution
9. *1 M TCEP:* 2.86 mg of TCEP weighed and made up to 1 mL solution.

5.2.3 Prepation of Sypro Orange Dye (50×)

6μL 5000× sypro orange stock added with 54μL of milliQ water to prepare 500×. 60μL of 500× stock added to 540μL of milliQ water to give the final 50× stock (see **Note 1**).

5.3 Methods

5.3.1 Priming TSA

1. Protein samples were centrifuged at 12,000 rpm for 10 min at 4 °C to check for sedimentation.
2. Collect supernatant, quantify using nanodrop.
3. Different protein concentrations (0.5, 1, 2, 5, 10μM) were evaluated in TSA using protein storage buffer.
4. Based on the initial results with qualitative melt curves, a single protein concentration of 2μM was selected and further evaluated at different pH conditions. All experiments were carried out in triplicates.

5.3.2 Procedure in Setting up TSA Experiment

1. All experimental procedure was done at 4 °C using ice, unless and otherwise stated to be at different temperature.
2. Positive control: Only protein of interest in its storage buffer.
3. Negative control: Only buffer (see **Note 2**).
4. Reference control: Standard protein with known Tm in TSA experiment (see **Note 3**).
5. 125μL of 12 mM sodium acetate buffer was added to their respective vials as marked (Table 5.1) (see **Note 4**)
6. 6μL of protein of interest was added in all vials except negative control. (see **Note 5**)
7. The content was mixed thoroughly with a pipette using a simple aspiration and release method. Non-mixing of solution results in absence of a proper experiment run.

Table 5.1 Details of preparation of experiment vials

	Concentrations		Final volume (μL)			
Reagent	Stock	Final working	pH 4.5	pH 5.0	pH 5.5	pH 6.0
Sodium acetate	12 mM	10 mM	125	125	125	125
Protein	50μM	2μM	6	6	6	6
Sypro orange	50×	5×	15	15	15	15
MilliQ water			4	4	4	4
Total volume			150	150	150	150

Table 5.2 Details of preparation of control vials

	Concentrations		Final volume added (μl)		
Reagents	Stock	Final working	Positive control	Negative control	Reference control
Protein	50μM	2μM	6	0	6
Sypro orange	50×	5×	15	15	15
Protein storage buffer			129	135	129
Total volume			150	150	150

8. Ensure separate reference control vial planned was pipetted out with *reference protein* prepared in storage buffer (Tables 5.1 and 5.2).
9. 15μL of Sypro Orange dye was added to all vials including controls. (see **Note 6)**
10. Seal the plate using a sealant. Details of the individual well are shown in Fig. 5.1.
11. Load the plate onto CFX96™ Real-Time PCR Detection system—C1000 Thermal Cycler.
12. Enter the plate details based on the experiment plan.
13. Run the experiment.
14. The samples heated from 15 °C to 95 °C with an increment of 1 °C per 30 s.

5.3.3 Analysis

Experiment Condition	Tm (°C)	Remarks
2μM target protein in storage buffer	42	
2μM target protein +10 mM sodium acetate pH 4.5	. . .	No proper melt curve recorded
2μM target protein +10 mM sodium acetate pH 5.0	30	
2μM target protein +10 mM sodium acetate pH 5.5	36	
2μM target protein +10 mM sodium acetate pH 6.0	36	

Target protein in storage buffer (control) recorded a Tm ~ 42 °C (Fig. 5.2). However, Tm of the target protein in Sodium acetate buffers was recorded between 30 °C and 36 °C. Among the buffers evaluated, pH 5.5 and 6.0 were relatively closer

→	↓	↓		→	↓	↓	
			Conditions				Conditions
A1	A2	A3	2uM Target Protein + 10mM NaoAC pH 4.5	A4	A5	A6	Negative Control -1 10mM NaoAC pH 4.5
B1	B2	B3	2uM Target Protein + 10mM NaoAC pH 5.0	B4	B5	B6	Negative Control - 2 10mM NaoAC pH 5.0
C1	C2	C3	2uM Target Protein + 10mM NaoAC pH 5.5	C4	C5	C6	Negative Control - 3 10mM NaoAC pH 5.5
D1	D2	D3	2uM Target Protein + 10mM NaoAC pH 6.0	D4	D5	D6	Negative Control - 4 10mM NaoAC pH 6.0
E1	E2	E3	2uM Target Protein + Target protein storage buffer	E4	E5	E6	Negative Control - 5 Target protein storage buffer
F1	F2	F3	2uM P38a - Refernce Protein + Reference protein Storage buffer	F4	F5	F6	
G1	G2	G3		G4	G5	G6	
H1	H2	H3		H4	H5	H6	

Fig. 5.1 Plate set up for buffer evaulation (Source: Jubilant Biosys)

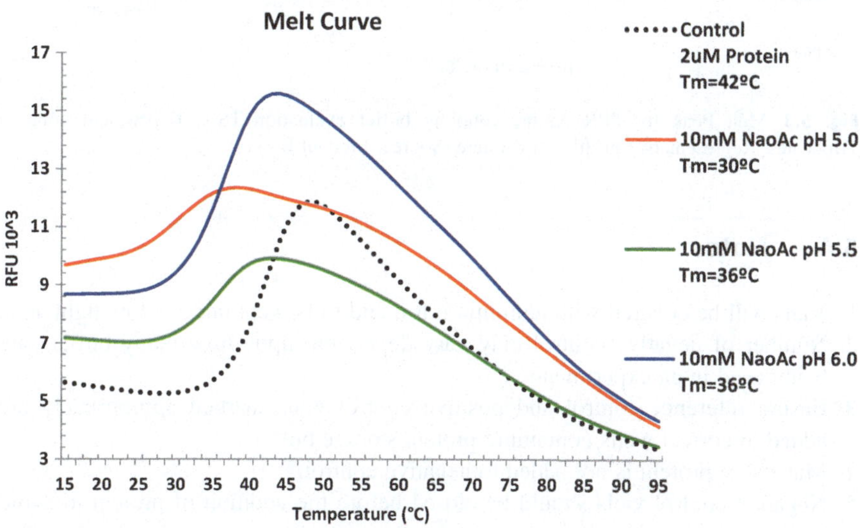

Fig. 5.2 Melt Curve for SPR Amine coupling buffer evaluation TSA. Experiment done in Triplicates. Representative profile shown here (Source: Jubilant Biosys)

to control. Buffer containing pH 5.5 was used to immobilize proteins onto the SPR sensor (Table 5.3 and Fig. 5.3).

Table 5.3 Troubleshooting guide for Thermal Shift Assay

Problem	Cause	Action
Low RFUs	Sypro stock might have been exposed to light/quenched already	• Prepare fresh stock with entire handling in low light conditions • Also, procure fresh Sypro orange reagent
	Protein concentration issue	• Reoptimize protein concentration to be used for TSA
No proper melt curve	Protein not stable at the TSA buffer conditions	• Check varying buffer conditions and select the suitable condition which records the proper melt curve for the given protein

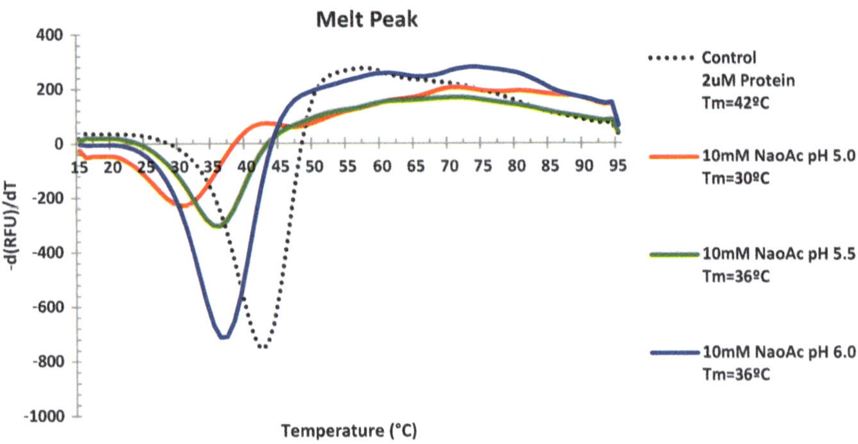

Fig. 5.3 Melt Peak for SPR Amine coupling buffer evaluation TSA. Experiment done in Triplicates. Representative profile shown here (Source: Jubilant Biosys)

5.4 Notes

1. Vials will be covered with aluminium foil and to be kept under a low light area.
2. Number of negative control may vary depending upon how many buffers are being used in an experiment.
3. Ensure reference control and positive control vials marked appropriately are added to correct wells containing protein storage buffer.
4. Make sure protein is not added to negative control.
5. Negative control vials should be closed before the addition of protein to avoid cross contamination.
6. Maintain distant dull light fair enough to pipette out.

References

1. Huynh K, Partch CL (2015) Analysis of protein stability and ligand interactions by thermal shift assay. Curr Protoc Protein Sci 79:28.9.1–28.9.14
2. Niesen FH, Berglund H, Vedadi M (2007) The use of differential scanning fluorimetry to detect ligand interactions that promote protein stability. Nat Protoc 9:2212–2221

Protein Stability Using Thermal Shift Assay (TSA): DMSO Tolerance

6

V. Swarnakumari and A. L. Theerthan

Abstract

SPR assay requires ligation of protein to biosensor chips by amine coupling method or biotin capture methods. However, prior to ligation of protein onto the sensors, protein of interest need to be evaluated for DMSO tolerability and stability. Buffer composition is an important aspect of SPR assay due to its high sensitivity for small changes in refractive indices. DMSO is one of the major factors that influence buffer mismatch leading to failure of SPR assays. TSA is used as a tool to evaluate the stability of proteins in different DMSO concentrations in the buffer based on the Melt curve and Tm recorded.

Keywords

TSA · Melt curve · Tm · Refractive index

6.1 Introduction

Differential Scanning Fluorimetry (DSF) commonly referred to as Thermal Shift Assay (TSA) is the most commonly used method to make a quick evaluation of protein stability based on the temperature at which protein melts in presence of fluorescent dye, Sypro orange [1]. The principle of TSA is with increasing temperature, proteins unfold, leading to exposure of hydrophobic amino acids present in proteins. The Sypro orange dye binds to these hydrophobic amino acid residues of protein and fluorescence is measured (Ex:492 nm/Em:610 nm). Increase in fluorescence is the direct measure of protein unfolding [2]. Melting Temperature (Tm) is the

V. Swarnakumari (✉) · A. L. Theerthan
Structural Biology Division, Discovery Biology, Jubilant Biosys Ltd., Bengaluru, Karnataka, India
e-mail: Swarnakumari.V@jubilantbiosys.com

S. M. Zaheer, R. Gosu (eds.), *Methods for Fragments Screening Using Surface Plasmon Resonance*, https://doi.org/10.1007/978-981-16-1536-8_6

temperature at which half of the protein is unfolded. The experiment involves measuring the change in Tm of the protein under varying buffering conditions in comparison with protein storage buffer [3]. Each protein is tested at different DMSO concentrations before starting an SPR assay. When screening a DMSO solubilized library in SPR, it is important to account for the effect of DMSO on protein stability. Some proteins tolerate DMSO poorly, demanding a lower concentration to be used. If DMSO concentration is well tolerated for a protein, higher concentration (up to 5% v/v) could be used [4]. It is important that the amount of DMSO is within the tolerance limit of the protein. To study the effect of various concentrations of DMSO on a protein of interest, series of DMSO concentrations ranging from 0% to 10% were tested by TSA. This chapter focuses on the development of TSA protocol to evaluate the optimal DMSO in running buffer.

6.2 Materials

6.2.1 Reagents Required for TSA

1. SYPRO™ Orange Protein Gel Stain—5000× Concentrate in DMSO, (ThermoFisher (Invitrogen)-S6650).
2. Multiplate™ Low-Profile 96-Well Unskirted PCR Plates (Bio Rad MLL9601).
3. Microseal "B" PCR Plate Sealing Film, adhesive, optical (Bio Rad—MSB1001).
4. Single channel pipettes—p2.5, p10, p20, p100 (Eppendorf).
5. CFX96™ Real-Time PCR Detection system—C1000 Thermal Cycler (Bio-Rad).
6. 100% DMSO (Sigma-34,869)..
7. HEPES (Sigma-H7006-500G).
8. NaCl (Sigma-S9888-500G).
9. 100% Glycerol (Sigma-G7893-500ML).
10. TCEP (Hampton-HR2-801).

6.2.2 Buffer and Reagent Preparation for TSA

1. *1 M HEPES:* 2.60 g of HEPES sodium salt weighed and dissolved in 8 mL of autoclaved milliQ water, pH adjusted to 7.7, and final volume made up to 10 mL.
2. *5 M NaCl:* 2.92 g weighed and dissolved in 10 mL of autoclaved milliQ water. Solution was filtered with 0.22 μm filter and stored at RT until use.
3. 1 M TCEP: 0.286 g weighed and dissolved in 1 mL of autoclaved milliQ water.

6.2.3 Prepation of Protein Storage Buffer (50 mL)

1 mL of 1 M HEPES (pH 7.7) (final conc. 50 mM), 5 mL of 5 M NaCl (final conc. 500 mM), 5 mL of 100% glycerol (final conc. 10%), 30 μL of 1 M TCEP (final conc. 0.5 mM) was added to 35 mL of milliQ water, stirred and the volume made up to 50 mL.

6.2.4 Prepation of Protein or Ligand Stock (50 uM)

15.5 μL of 54.93 μM protein was diluted with 1.5 μL of protein Storage buffer to get a final concentration of 50 μM.

6.2.5 Prepation of Sypro Orange Dye (50×)

6 μL 5000× Sypro Orange stock added with 54 μL of milliQ water to prepare 500×. 60 μL of 500× stock was added to 540 μL of milliQ water to give the final 50× stock. (see **Note 1**).

6.3 Methods

6.3.1 Priming TSA

1. Protein samples were centrifuged at 12000 rpm for 10 min at 4 °C to check for sedimentation.
2. Supernatant was collected and quantified using nanodrop.
3. Different protein concentrations (0.5, 1, 2, 5, 10 μM) were evaluated in TSA using a protein storage buffer.
4. Based on the initial results with qualitative melt curves, single protein concentration of 1 μM was selected and further evaluated at different DMSO concentrations. All experiments were carried out in triplicates.

6.3.2 Procedure for Setting Up TSA Experiment

1. All experimental procedure was done at 4 °C using ice, unless and otherwise stated to be at different temperature.
2. Positive control: Protein of interest in its storage buffer.
3. Negative control: Only buffer (see **Note 2**).
4. Reference control: Standard protein with known Tm in TSA experiment *(see* ***Note 3****)*
5. Pipette out respective buffer solutions first to the triplicate vials as per markings in ice (Table 6.1).

Table 6.1 Outlines the amount of volume required for each component

	G1	G4	G7	G10	H1	H4	H7
	1% DMSO	2% DMSO	5% DMSO	10% DMSO	Positive control	Negative control	Refernce control
Components	Volume in (μl)						
Protein storage buffer	130.5	129	124.5	117	132	135	120
50 μM protein	3	3	3	3	3	0	15
100% DMSO	**1.5**	**3**	**7.5**	**15**	**0**	**0**	**0**
50× Sypro orange dye	15	15	15	15	15	15	15
Total volume	150	150	150	150	150	150	150

6. 3 μL of protein was added to their respective vials as marked (Table 6.1) (see **Note 4**): Make sure protein is not added to negative control)
7. The content was mixed thoroughly with a pipette with a simple aspiration and release method (see **Note 5**).
8. Different concentration of DMSO was added to respective wells as shown in Table 6.1. (see **Note 6**).
9. 15 μL of Sypro Orange dye (50×) was added to all vials. (see **Note 7**) Maintain distant dull light fair enough to pipette out the required dye. Presence of direct light will bleach dye.
10. Finally, mix the content (total volume 150 μL) and transfer 50 μL from vial to TSA plate to their corresponding wells.
11. Plate was sealed with microseal "B" PCR Plate Sealing Film.
12. Plate was loaded onto CFX96™ Real-Time PCR Detection system—C1000 Thermal Cycler.
13. Open melt curve protocol from the software as shown in Fig. 6.1.
    (a) The protocol: Samples heated from 15 °C to 95 °C with an increment of 1 °C per 30 s.
14. Plate details were loaded based on the experimental plan (Fig. 6.2) (see **Note 8**).
15. Experiment was started by selecting "Start Run" tab (Fig. 6.3).

6.3.3 Results and Analysis

Target protein tolerates DMSO in all tested concentrations (Fig. 6.4). However, at 5% and 10% DMSO concentration, initial RFU were slightly higher and negative Tm shift of 2–3 °C was recorded (Fig. 6.5). Based on the above information, 2% DMSO concentration (optimal) in running buffer was fixed and used in subsequent screening experiments (Table 6.2).

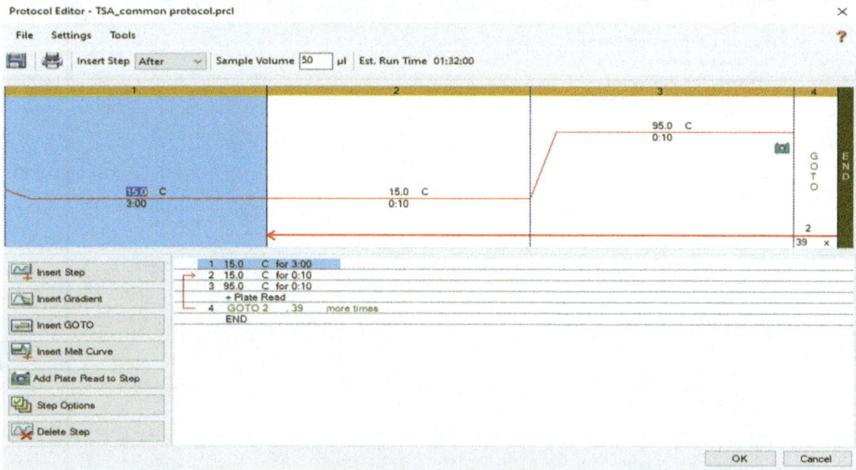

Fig. 6.1 DMSO tolerance evaluation in TSA. Melt curve protocol (Source: Jubilant Biosys)

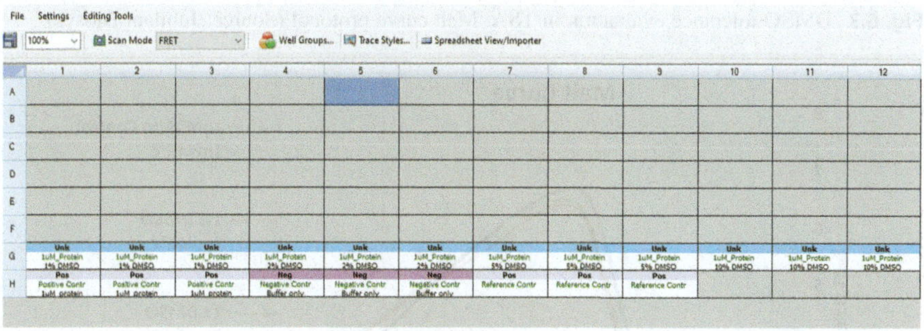

Fig. 6.2 DMSO tolerance evaluation in TSA. Details of plate file (Source: Jubilant Biosys)

Condition	Tm (°C)	Remarks
Positive control (Apo)	46	
1% DMSO	46	Melt curve as equal to Apo protein curve
2% DMSO	45	Melt curve as equal to Apo protein curve, with 1 °C less Tm than Apo protein TM
5% DMSO	44	Slightly higher initial baseline than Apo protein curve with 2 °C negative Tm.
10% DMSO	43	Higher initial baseline with 3 °C negative Tm than Apo protein

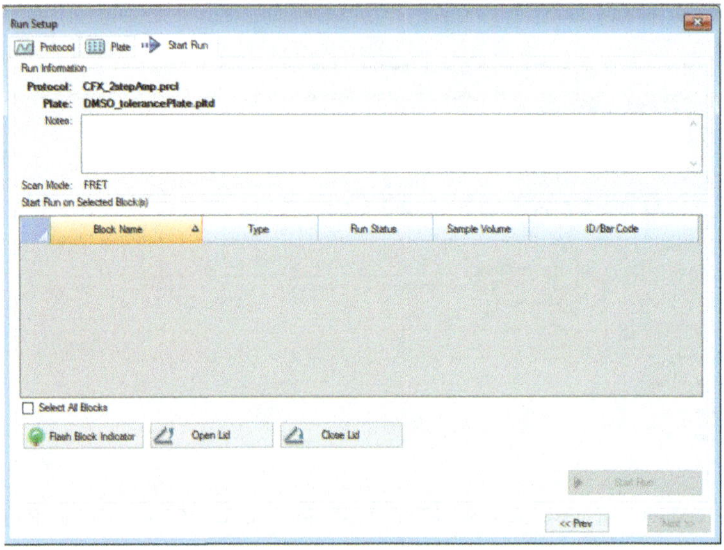

Fig. 6.3 DMSO tolerance evaluation in TSA. Melt curve protocol (Source: Jubilant Biosys)

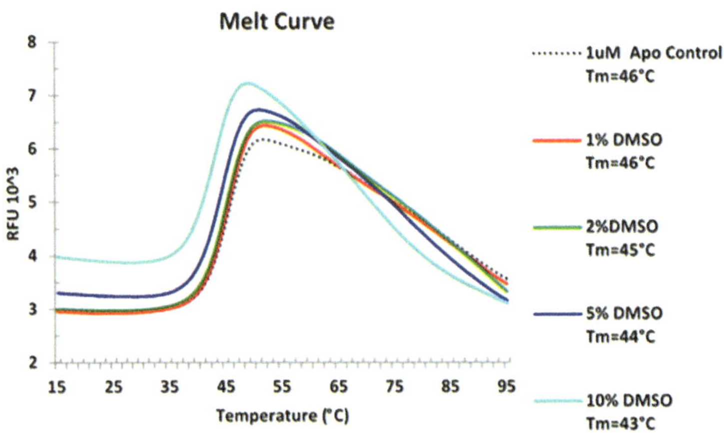

Fig. 6.4 Melt curve for SPR assay—DMSO evaluation in TSA. Experiment done in triplicates. Representative profile shown here (Source: Jubilant Biosys)

6.4 Notes

1. Vials were covered with aluminium foil and kept under a low light area to avoid fluorescence bleach.
2. Number of negative control may vary depending upon the different buffers used in an experiment.

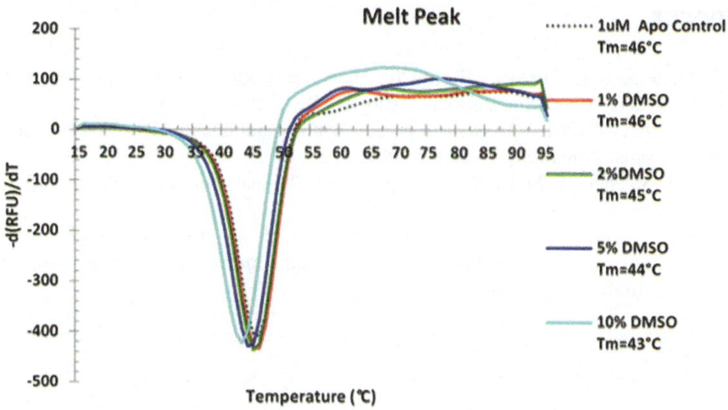

Fig. 6.5 Melt peak for SPR assay—DMSO evaluation in TSA. Experiment done in triplicates. Representative profile shown here (Source: Jubilant Biosys)

Table 6.2 Troubleshooting guide for TSA

Problem	Cause	Action
High initial RFU	• Destabilisation of protein (unfolded and folded) or aggregation	• Reduced concentration of components that denatures the protein • Centrifuge protein before adding
	• Detergents	• Detergents should be avoided as these will interfere with Sypro Orange
Triplicate curve mismatch	• Air bubble in the well	• Precaution to be taken while pipetting the solution to the well
	• Mismatch in Volume	• All triplicate volume to be equally dispensed

3. Ensure reference control and positive control vials marked appropriately are added to correct wells containing protein storage buffer.
4. Make sure protein is not added to negative control.
5. Non-mixing of solution results in absence of a proper experiment run.
6. Do not add DMSO solution to positive control, negative control, and reference control.
7. Maintain distant dull light fair enough to pipette out the required dye. Presence of direct light will bleach dye.
8. The plate map and samples well should match, failing which will give incorrect results.

References

1. Gao K, Oerlemans R, Groves MR (2020) Theory and applications of differential scanning fluorimetry in early-stage drug discovery. Biophys Rev 12(1):85–104
2. Huynh K, Partch CL (2015) Analysis of protein stability and ligand interactions by thermal shift assay. Curr Protoc Protein Sci 79:28.9.1–28.9.14
3. Redhead M, Satchell R, McCarthy C et al (2017) Thermal shift as an entropy-driven effect. Biochemistry 56(47):6187–6199
4. Nettleship JE, Brown J, Groves MR et al (2008) Methods for protein characterization by mass spectrometry, thermal shift (ThermoFluor) assay, and multiangle or static light scattering. Methods Mol Biol 426:299–318

Part III

Fragment Screening Using SPR

7 Selection and Identification of Fragment Library

Rajendra Kristam and Ramachandraiah Gosu

Abstract

In this chapter, we provide details of how to proceed for the selection of fragments based on various criteria. Smart selection of diverse fragments is essential for starting fragment-based drug discovery efforts as multiple libraries are available commercially built on different clustering criteria, which are redundant to an extent. Although many criteria can be applied to select a diverse set of fragments toward specific targets, we describe applications of criteria that in general result in the selection of a diverse set of fragments keeping in mind their later expansion into full small molecules and their development into possible drugs with good profiles.

Keywords

Fragment-based drug discovery (FBDD)

7.1 Introduction

Several chemical vendors supply a collection of fragments, both general fragment as well as focused fragment collections, such as amine fragments or acid-containing fragment collections. The description here is for a general collection of fragments

R. Kristam (✉)
Computational Chemistry Division, Medicinal Chemistry, Jubilant Biosys Ltd., Bengaluru, Karnataka, India
e-mail: Rajendra.Kristam@jubilantbiosys.com

R. Gosu
Structural Biology Division, Discovery Biology, Jubilant Biosys Ltd., Bengaluru, Karnataka, India
e-mail: ramachandraiah.gosu@jubilanttx.com

S. M. Zaheer, R. Gosu (eds.), *Methods for Fragments Screening Using Surface Plasmon Resonance*, https://doi.org/10.1007/978-981-16-1536-8_7

and not any focused set. Forty-four fragment libraries were collected from about 15 chemical vendors, with the sizes of libraries ranging from a low as 480 fragments from Prestwick's core set to 110,557 fragments from Enamine. Since the collection of fragments can have fragments with moieties that could be frequent hits among in vitro assays, several filters were applied to remove such fragments. REOS and PAINS filters were applied to trim the fragment collections by filtering off the collections of reactive compounds and moieties that are among the frequent hitting compounds. Various physicochemical properties (MW, AlogP, PSA, #HBA, #HBD, LogS, and #rings) were calculated for all the libraries, along with a "natural-product-likeness" score and Fsp^3 (fraction of sp^3 centers). Natural product-like score represents the fraction of the fragment representing natural products. Natural products are of particular importance because they were optimized in the course of evolution to have constructive and effective interactions with most biological receptors. They are therefore considered good starting point for designing new drugs [1, 2]. For all these properties, the average, min/max, skewness, and kurtosis were calculated. These calculations were carried out using Schrodinger/Canvas module.

All the libraries were then clustered, using atom-pair fingerprints at a Tanimoto coefficient of 0.7 similarity. Focus was on the number of clusters, sizes of clusters, and the number of singletons in each library. All these indirectly represent the diversity of the library. The choice of a Tanimoto coefficient of 0.7 is arbitrary, but it can represent a threshold with fragments clustered in such a way that the resulting clustered collection will be having neither very similar fragments nor very diverse fragments. Clustering was carried out using the Leader-Follower clustering method as well as the Sphere-exclusion method as implemented in Schrodinger/Canvas module.

In the next step, the top 14 libraries in terms of (a) number of clusters as a percentage of library size, (b) number of singletons as a percentage of library size, (c) natural-product likeness score, (d) Fsp3 [3, 4], and (e) size of the library. The libraries for this listing were chosen so that about 2000 fragments are present in the libraries after all filters were applied.

Listing of the top 14 fragment libraries, with a library size of around 2000 fragments. #cpds: number of fragments, reduced set: the total number of fragments after application of all filters, %rem: percentage of the original library removed due to application of filters, Avg.Fsp3: average Fsp3 value, % NPlike: percentage of library size having natural product-like fragments, % single: percentage of library size comprising singletons, and % clusters: percentage of library size comprising clusters.

Among the listed libraries, the top six were chosen based on the following criteria

(a) size and diversity of the library for direct purchase,
(b) natural-product likeness score and,
(c) Fsp3.

Table 7.1 Top 14 fragment libraries selected for SPR screening

S. No	Library	# cpds	Reduced set	% remaining	Avg. Fsp3	% NPlike	% single	% clusters
1	UORSY Fragments with solubility	936	923	1%	0.44	66%	5.31	23.73
2	Selleck L1600 Fragment Library 1015	1015	870	14%	0.22	58%	2.64	11.61
3	OTAVA Solubility Fragment Library	1021	981	4%	0.24	76%	2.85	14.27
4	BIONET 2nd Gen Premium Fragment Library	1221	1139	7%	0.26	66%	2.29	13.05
5	Compound Cloud Inventory Selcia	1366	1137	17%	0.24	64%	4.22	15.39
6	LC 3D Fragment Library	1377	1283	7%	0.64	76%	2.26	15.12
7	Chemspace Tiny Fragments SDF	1418	1393	2%	0.61	3%	0.72	9.83
8	Prestwick Drug FragLib List	1460	1326	9%	0.28	74%	3.09	15.23
9	Prestwick Drug Frag 1456	1700	1551	9%	0.3	64%	2.13	12.83
10	LC Low MW Fragment Library Advanced Set	1753	1675	4%	0.45	82%	2.15	12.9
11	Chemspace Singleton Fragments SDF	2096	2049	2%	0.6	78%	4.05	18.89
12	OTAVA Chelator Fragment Library	2441	2230	9%	0.23	74%	1.3	7.35
13	OTAVA Stereogenic Centers Fragment Library	2449	2290	6%	0.45	79%	1.83	9
14	Chemspace Spiro Fragments SDF	2466	2344	5%	0.75	82%	0.73	6.53

The libraries highlighted in green in Table 7.1 represent the most diverse libraries after the application of various filters and considering all calculated values. The libraries with the least number of fragments removed due to the application of all filters are highlighted in the fifth column. The libraries with most fragments showing a higher Fsp3 value are highlighted in the sixth column, while those with a higher natural product-like score are highlighted in the seventh column. The eighth column represents the percentage of library size comprising singletons after clustering, while the ninth column represents the percentage of library size comprising clusters after clustering. All these parameters were considered in selecting the top six fragment libraries as highlighted in the second column. The workflow below represents the flow of steps for the selection of the top-ranking diverse libraries (Fig. 7.1).

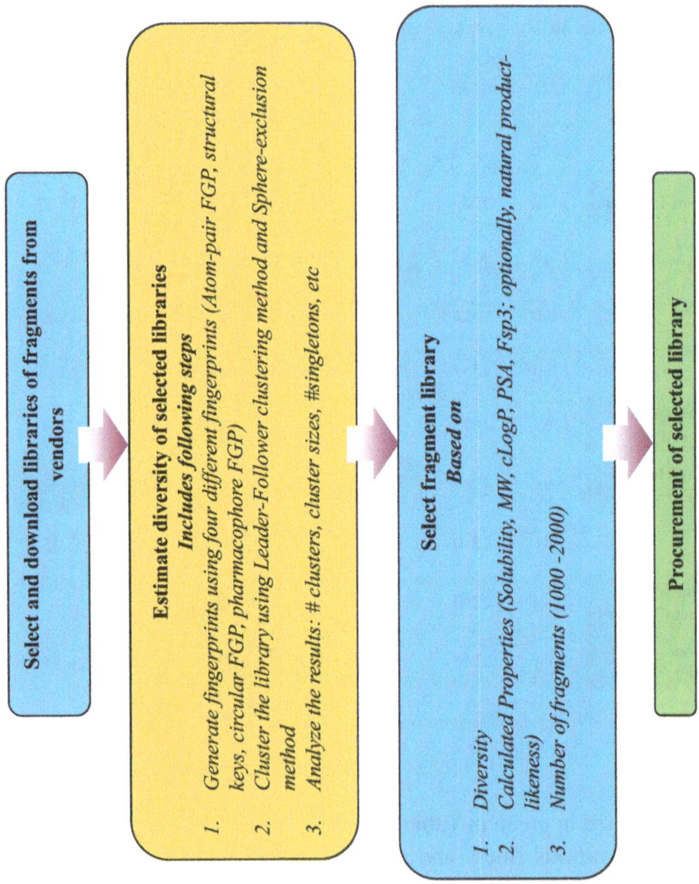

Fig. 7.1 Flow scheme in selection of diverse fragments library (Source: Jubilant Biosys)

References

1. Ertl P, Roggo S, Schuffenhauer A (2008) Natural product-likeness score and its application for prioritization of compound libraries. J Chem Inf Model 48(1):68–74
2. Jayaseelan KV, Moreno P, Truszkowski A et al (2012) Natural product-likeness score revisited: an open-source, open-data implementation. BMC Bioinfo 13:106
3. Wei W, Cherukupalli S, Jing L et al (2020) Fsp3: a new parameter for drug-likeness. Drug Discov Today S1359-6446(20):30297-X. https://doi.org/10.1016/j.drudis.2020.07.017
4. Lovering F, Bikker J, Humblet C (2009) Escape from flatland: increasing saturation as an approach to improving clinical success. J Med Chem 52(21):6752–6756

Preparation of SPR Sensor

8

Krishnakumar Vaithilingam and Sameer Mahmood Zaheer

Abstract

Preparation of SPR sensor is an important part, prior to assay and screening. Biosensors are highly fragile and form the basic component of the SPR instrument. Improper sensitization or activation of biosensors can lead to nonuniform ligation of ligands, which result in variations between triplicates during SPR assays. Inadequate activation is also responsible for the ligand drift post immobilization. This chapter provides information to prepare SPR sensor and immobilization of ligand of interest.

Keywords

HTS · Biotin–Neutravidin · Flow cells · Gold surface · Amine coupling

8.1 Introduction

Biosensors (Table 8.1) are coated with a biocompatible matrix, which has to be uniform and non-denaturing [1]. These components help in the reduction of nonspecific binding. The matrix is coated on a gold surface that is chemically inert, biocompatible, has excellent conductivity, and a high surface to volume ratio [2]. PioneerFE SPR has three flow cells in a biosensor (Fig. 8.1a), first flow cell

K. Vaithilingam (✉) · S. M. Zaheer
Structural Biology Division, Discovery Biology, Jubilant Biosys Ltd., Bengaluru, Karnataka, India
e-mail: Krishnakumar.V@jubilantbiosys.com; Sameer.Mahmood@jubilantbiosys.com

S. M. Zaheer, R. Gosu (eds.), *Methods for Fragments Screening Using Surface Plasmon Resonance*, https://doi.org/10.1007/978-981-16-1536-8_8

Table 8.1 Details of biosensors used for high through put screening

Biosensor	Surface chemistry	Immobilization method	Recommended use	Advantages
CDH (COOHV) (Cat. No: 19-0128)	Carboxymethyl dextran three-dimensional hydrogel surface with carboxylic acid	Amine coupling	Small molecule–ligand interaction, fragment screening, targets with low binding activity	• Target immobilization without derivatization or tags • High capacity dextran suitable for immobilizing a large amount of ligand • Biocompatible with a wide range of molecules • Forms highly stable covalent bonds • Effective over a wide range of pH
SADH (BioCap) (Cat. No: 19-0130)	Streptavidin immobilized in three-dimensional carboxymethyl dextran hydrogel	Capture via biotin tag	Intermediate (1–25 kDa) molecule kinetics with biotinylated ligands	• Highly efficient capture in a wide range of pH • Only less quantity of ligand required • Single step immobilization • Surface has a lower electrostatic charge compared to amine coupling sensors

was used for ligand ligation, second flow cell was used as a reference (no ligand loaded), and third flow cell was reserved as a backup. Several biosensors with different surface chemistry and immobilization method were available for HTS screening. We selected amine coupling and biotin–neutravidin capture techniques for our HTS screening (Table 8.2).

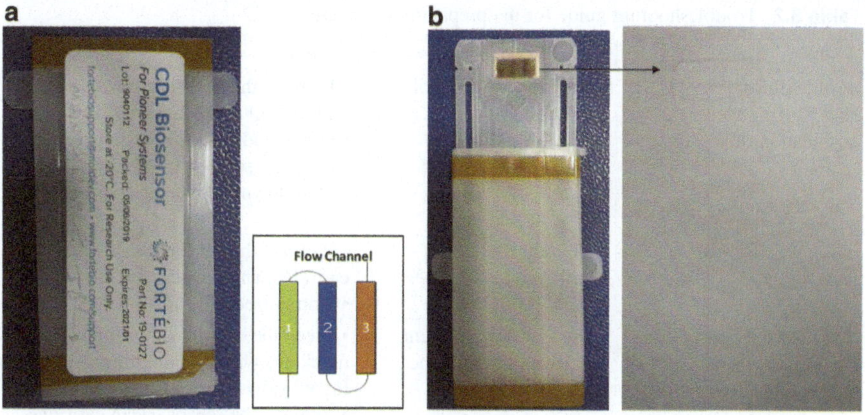

Fig. 8.1 (**a**) PioneerFE SPR sensors has three flow channels (FC-1-2-3). The ligands can be loaded from either 1–2-3 or 3–2-1 sequence. 1 RU corresponds to 1 pg/mm^2. Ligands are loaded in channel 1 or channel 3, with channel 2 is dedicated as reference channel. The reference channels assist in double subtraction in analysis. (**b**) PioneerFE SPR sensors cleaning cartridge with impression of flow cell after cleaning (arrow) (Source: Jubilant Biosys)

8.2 Materials

8.2.1 Immobilization Buffer

10 mM Sodium acetate: 0.34 g of Sodium acetate (Cat. No: S8625, Sigma Aldrich), was added into 250 mL of milliQ water and pH was adjusted to 5.5.

8.2.2 Preparation of 0.4 M EDC

153 mg of EDC (Cat. No: E1769, Sigma Aldrich) was added to 2 mL of MilliQ water and mixed well.

8.2.3 Preparation of 0.1 M NHS

23 mg of NHS (Cat. No: 130672, Sigma Aldrich) was added to 2 mL of milliQ water and mixed well (see **Note 1**).

8.2.4 Preparation of Ligand for Immobilization

- 50μg/mL ligand was prepared in 500μL of immobilization buffer.

Table 8.2 Troubleshooting guide for the preparation of sensor

Problem	Cause	Action
Profile dips fall <90% and >110% after installation of sensor	• Improper docking of sensor	• Remove the sensor, clean cartridge with milliQ water and isopropyl alcohol, and reinstall the sensor • Normalized with 100% DMSO • Shut down the machine system and restart
	• Non-equilibrium of profile dips	• Remove the sensor, clean with cartridge, and reinstall the sensor • Normalized with 100% DMSO
Irregular dips post clean	• Dust particles present onto the surface of the sensor flow cells	• Clean the sensor with milliQ water, quick dry with air pump immediately (avoid slow drying) • Remove the sensor, clean cartridge with milliQ water and isopropyl alcohol, and reinstall the sensor
Buffer leakage from stroke syringe pump	• Back pressure in pump	• Alternate 10× priming with milliQ water and 0.4% NaOCl and this is repeated 5 times • Rigorous cleaning with 10% DMSO and 0.4% NaOCl for 16 h • Followed by 10× priming twice with milliQ water to remove NaOCl from the tubes
Ligand stabilization and drift in RU	• Running buffer	• Change to high salt concentration in running buffer • Check pH of running buffer • Change running buffer
Low RUs	• High flow rate	• Reduce the flow rate
	• Low protein conc.	• Increase protein conc.
	• Running buffer pH and protein pI mismatch	• pH should be closer to the pI of the protein
	• Activation reagents	• Prepare fresh activation reagents (EDC and NHS) from fresh stock
Nonuniform activation of flow channels	• Variation of flow rate between different channels	• Forward injection FC-1-2-3 and reverse injection FC-3-2-1 is required to get uniformity

8.2.5 Preparation of Blocking Reagent

0.5 M Ethanolamine: 200μL of 1 M stock Ethanolamine (Cat. No: 411000, Sigma Aldrich, pH 8.5) was added to 200μL of immobilization buffer and stored at 4 °C until use.

8.2.6 Preparation of of 20% DMSO

20 mL of 100% DMSO (Cat No: 34869, Sigma Aldrich) dissolved in 80 mL of milliQ water.

8.2.7 Preparation of of 10 mM DTT

0.15 g of DTT (Cat. No: D9779, Sigma Aldrich) dissolved in 100 mL of milliQ water.

8.2.8 Preparation of of 0.05% Tween20

50μL of 100% Tween20 (Cat. No: P1379, Sigma Aldrich) dissolved in 99.5 mL of milliQ water.

8.2.9 Preparation of 1000 mL Running Buffer (1× PBS)

100 mL of 10 X PBS (Cat. No: 7011-044, Gibco), 100 mL of 20% DMSO, 100 mL of 10 mM DTT, 100 mL of 0.05% Tween20 was added. pH was adjusted to 7.4, and make upto 1000 mL with milliQ water.

8.2.10 Maintenance Sensor

Maintenance sensor is used for calibration, quality check, and system flow path and cleaning sensor.

8.2.11 Cleaning Cartridge

Cleaning cartridge (Cat. No: PS99AFB, Fortebio) was used to remove dust particles if present in flow cells. It was used with deionized water and 70% ethanol or isopropyl alcohol.

8.3 Methods

8.3.1 Preinstallation Procedure

1. The biosensor (stored at −30 °C) was allowed to equilibrate at room temperature for 30 min prior to opening the cassette to prevent condensation on the sensor surface.

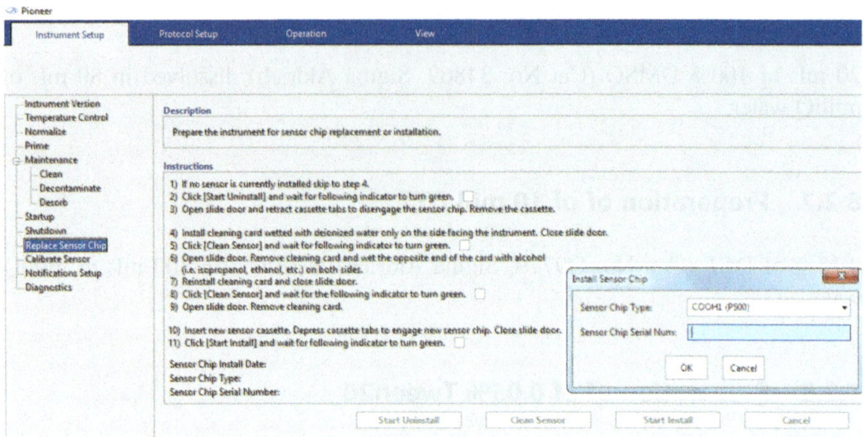

Fig. 8.2 Installation of Sensor with details (Source: Jubilant Biosys)

2. The flow cells were cleaned using new cleaning cartridge first with deionized water (wet one side of the cartridge facing the sensor), followed by 70% ethanol or isopropyl alcohol (wet both the sensor side) (Fig 8.1b) (see **Note 2**).
3. The equilibrated new sensor (CDH or SADH) was installed as per instructions on the PioneerFE software (8.2). (see **Note 3**).
4. All the details—sensor name, serial number was entered before installation (Fig. 8.2) (see **Note 4**)*:*
5. A good optical signal during the equilibration period showed overlayed profiles on all three optical channels and fall under 90–110% on the Y axis (Fig. 8.4a).
6. Once proper optical signal was observed after equilibration no further action was required. Priming was done with immobilization buffer four times to remove any sensor bound contaminants.
7. Post this sensor was ready for activation and immobilization of ligand of interest.

8.3.2 CDH Sensor

8.3.2.1 Activation of Sensor

1. *Activation buffer:* Equal volumes, 150μL of 0.4 M EDC and 150μL of 0.1 M NHS were mixed to get a final working conc. of 200 mM of EDC and 50 mM of NHS respectively. (see **Note 5**).
2. The 230μL of activation buffer was injected with a flow rate of 10μL/min to all the flow cells (FC-1-2-3).
3. 300 s dissociation was initiated to remove traces of contaminants present if any. (Fig. 8.3).

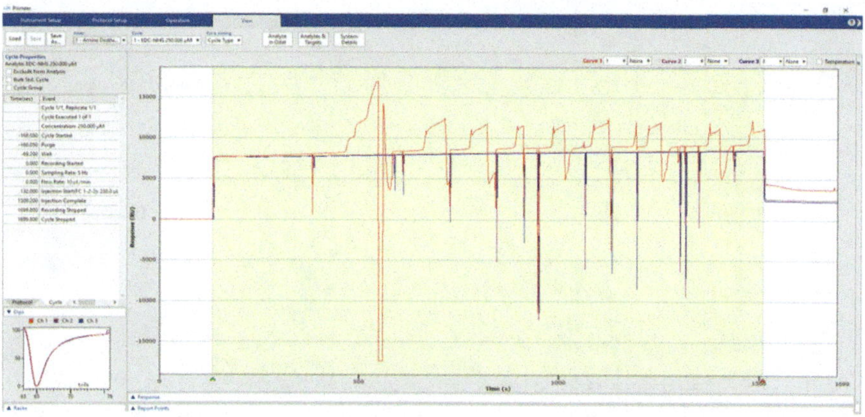

Fig. 8.3 Activation of Biosensor with EDC/NHS (Source: Jubilant Biosys)

8.3.2.2 Loading of Ligand onto the Sensor

1. 220μL of ligand (50μg/mL) was injected to flow channel 1 only, at a flow rate of 10μL/min followed by 300 s of dissociation (see **Note 6**)
2. Channel 2 was used as a reference channel (see **Note 7**).
3. 110μL of ligand 50μg/mL was injected to flow channel 3 at the flow rate of 10μL/min followed by 300 s of dissociation. (see **Note 8**).
4. The shift in dips observed in different channels confirms ligand loading (Fig. 8.4b).
5. The ligand of interest is loaded until the required RU (based on ligand) or saturation is reached (Fig. 8.5).

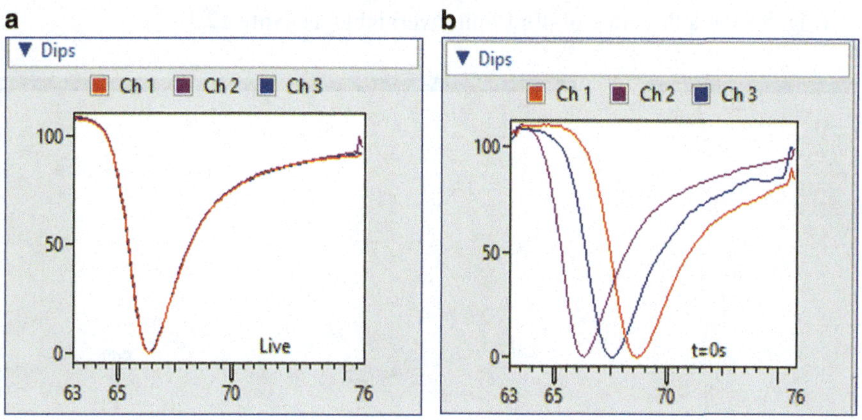

Fig. 8.4 (**a**) Dip profile of flow channel (FC-1-2-3) after installation of sensor before protein loading. (**b**) Dip profile of flow channel (FC-1-2-3) after ligand loading. Channel 1 (Ch1 is loaded with protein of interest at 50 ug/mL) Channel 2 (Ch2 is the reference), Channel 3 (Ch3 is loaded with protein of interest as back up) (Source: Jubilant Biosys)

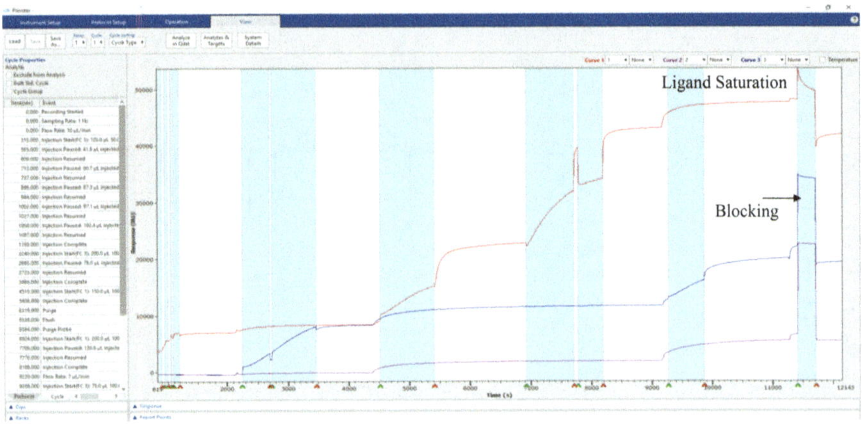

Fig. 8.5 Profile during ligand loading. Loading of ligand is stopped as the saturation threshold is reached. This followed by Blocking of sensor flow cells with Ethanolamine (Source: Jubilant Biosys)

8.3.2.3 Blocking of Ligand onto Sensor

1. 150μL of 1 M Ethanolamine was injected into all the flow channels (FC-1-2-3).
2. A flow rate of 30μL/min and a dissociation time of 300 s was maintained to complete the blocking process (Fig. 8.5) (see **Note 9**):

8.3.2.4 Stabilization of Ligand onto Sensor

1. The activated and ligand loaded sensor was maintained overnight in running buffer for stabilization (see **Note 10**).
2. We used PBS and HEPES buffers for most of our SPR assays. (see **Note 11**).
3. Stabilization process was initiated by priming the sensor with PBS running buffer (Fig. 8.6) at a flow rate of 40μL/min overnight (see **Note 12**).

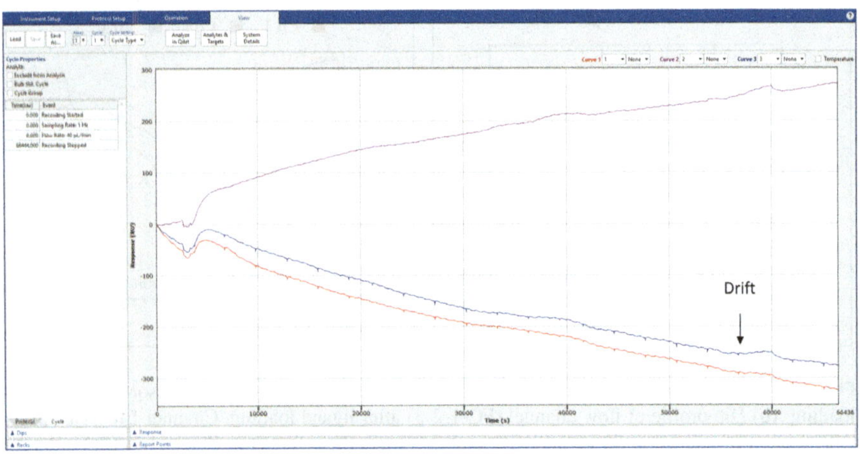

Fig. 8.6 Drift observed when the PBS buffer was used post ligand loading and blocking during stabilization process (Source: Jubilant Biosys)

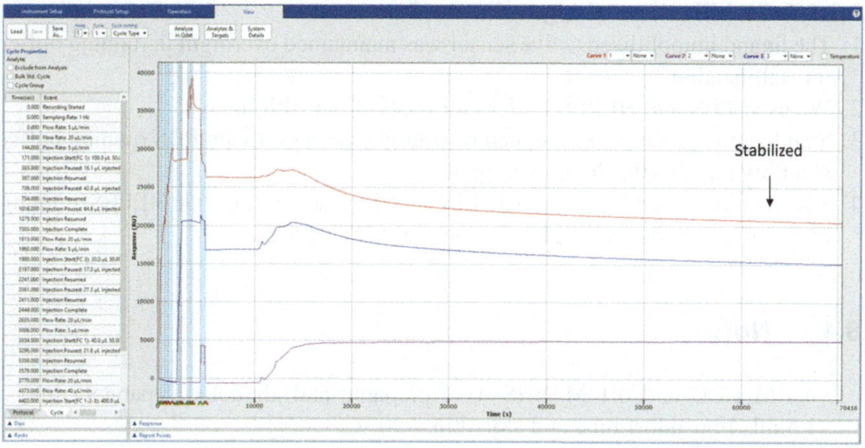

Fig. 8.7 Stabilization of protein onto the sensor observed when HEPES buffer was used (Source: Jubilant Biosys)

4. Stabilization process was initiated by priming the sensor with HEPES running buffer (Fig. 8.7) at a flow rate of 40μL/min overnight. (see **Note 13**).

8.3.3 SADH Sensor

8.3.3.1 Equilibration of Sensor

1. Equilibration: PBS was used for stabilistaion of SADH sensor.
2. The 230μL of PBS buffer was injected with a flow rate of 10μL/min to all the flow cells (FC-1-2-3).
3. 300 s dissociation was initiated to remove traces of contaminants present if any.

8.3.3.2 Loading of Ligand onto the Sensor

1. 220μL of biotinylated ligand (50–100μg/mL) was injected to flow channel 1 only, at a flow rate of 10μL/min followed by 300 s of dissociation
2. Channel 2 was used as reference channel.
3. 110μL of ligand 50μg/mL was injected to flow channel 3 at the flow rate of 10μL/min followed by 300 s of dissociation.

8.3.3.3 Blocking of Ligand onto Sensor

1. 150μL of 0.4 mM Biotin was injected into all the flow channels (FC-1-2-3).
2. A flow rate of 30μL/min and a dissociation time of 300 s was maintained to complete the blocking process.

8.3.3.4 Stabilization of Ligand onto Sensor

1. The ligand immobilized active sensor was maintained overnight in running buffer for stabilization.
2. We used PBS and HEPES buffers for most of our SPR assays.
3. Stabilization process was initiated by priming the sensor with PBS running buffer at a flow rate of 40μL/min overnight.
4. Stabilization process was initiated by priming the sensor with HEPES running buffer at a flow rate of 40μL/min overnight (Table 8.2).

8.4 Notes

1. 0.4 M EDC and 0.1 M NHS were prepared and aliquoted separately into a 1.5 mL Eppendorf tube and stored in −30 °C.
2. Cleaning was done until a visible impression of the flow cell was imprinted onto the cartridge.
3. Improper docking of the sensor leads to leakage of running buffer during the experiment run.
4. Dip profile was observed through "Dips window"in the Operation tab to inspect the quality of the optical signal prior to completion of sensor installation.
5. This cocktail has to be prepared freshly prior to activation of sensor.
6. Slower the flow rate, better the loading—Make sure the ligand is loaded only to flow channel 1 and not to flow channel 2.
7. Channel 2 was used for background correction while analysis..
8. Channel 3 is used as a backup.
9. Blocking helps in the prevention of nonspecific binding of analyte to the dextran surface.
10. Different running buffer were tested to check the stabilization process.
11. Ligand stabilization is a very important part during sensor preparation, as drifting (reduction in RUs) leads to variation in triplicates and binding affinities.
12. If a drift is observed during stabilization process, ligand loading procedure with HEPES buffer was reinitiated.
13. Screening cannot be performed until the sensor stabilizes.

References

1. Lei Z, Jian M, Li X et al (2020) Biosensors and bioassays for determination of matrix metalloproteinases: state of the art and recent advances. J Mater Chem B 8(16):3261–3291
2. Roussille L, Brotons G, Ballut L et al (2011) Surface characterization and efficiency of a matrix-free and flat carboxylated gold sensor chip for surface plasmon resonance (SPR). Anal Bioanal Chem 401(5):1601–1617

Optimization and Validation of Amine Coupling of Ligand

9

Krishnakumar Vaithilingam and Sameer Mahmood Zaheer

Abstract

In this chapter, we focus on the ideal strategy to optimize and validate amine coupling for SPR studies. In particular, we provide an in-depth description of each and every component required to establish validation before the screening of fragments. The primary emphasis is to maintain the protein in its natural state without losing its conformational change. The majority of proteins contain multiple amine groups so efficient immobilization of the protein is usually easily achieved without denaturation and inactivation of the protein. Optimization and validation protocol is presented here with important criteria and troubleshooting with action taken considered at each step.

Keywords

Immobilization · Amine coupling · RU · Sensogram

9.1 Introduction

Covalent immobilization of sensors is the primary and regularly utilized approach for linking the ligands to the surface [1]. Covalent immobilization is an additional strategy of choice for linking the molecules, for example, antibodies, proteins, and other biomolecules for affinity capture techniques. Of the three covalent immobilizing methods (Amine, Thiol, and Aldehyde), Amine coupling–surface activation using EDC/NHS complemented via coupling through primary amine groups (lysine residues) on the ligand is the most often used techniques used in

K. Vaithilingam (✉) · S. M. Zaheer
Structural Biology Division, Discovery Biology, Jubilant Biosys Ltd., Bengaluru, Karnataka, India
e-mail: Krishnakumar.V@jubilantbiosys.com; Sameer.Mahmood@jubilantbiosys.com

S. M. Zaheer, R. Gosu (eds.), *Methods for Fragments Screening Using Surface Plasmon Resonance*, https://doi.org/10.1007/978-981-16-1536-8_9

SPR [2]. The success of covalent linking of the ligand to the surface requires the ligand to preserve its global conformation upon immobilization to the surface and if necessary, that the protein be robust upon surface regeneration to eliminate bound analyte at the termination of the analysis cycle [3]. Amine coupling includes interactions between one or several of the multiple amine side chains on the ligand molecule and the activated surface [4]. However, amine coupling is highly random and as a consequence, the orientation of binding of the ligand cannot be controlled [5]. Therefore, a homogeneous orientation of the ligand is very important and considered key to the experiment this approach provides the most ideal result.

Several factors influence the effective linking of the ligand to the sensors, as these factors play a pivotal role in the success of the experiment. These factors include temperature of the ligand and machine, buffer pH, and charge mediated concentration of the ligand. This chapter provides a comprehensive detail to validate the protein immobilized onto the sensor using a known reference molecule.

9.2 Materials

9.2.1 30% DMSO Solution (100 mL)

30 mL of 100% DMSO (Cat No: 34869, Sigma Aldrich) dissolved in 70 mL of milliQ water.

9.2.2 20% DMSO Solution (100 mL)

20 mL of 100% DMSO (Cat No: 34869, Sigma Aldrich) dissolved in 80 mL of milliQ water.

9.2.3 10% DMSO Solution (100 mL)

10 mL of 100% DMSO (Cat No: 34869, Sigma Aldrich) dissolved in 90 mL of milliQ water.

9.2.4 10 mM DTT (100 mL)

0.15 g of DTT (Cat. No: D9779, Sigma Aldrich) dissolved in 100 mL of milliQ water.

9.2.5 0.05% Tween20 (100 mL)

50 µL of 100% Tween20 (Cat. No: P1379, Sigma Aldrich) dissolved in 99.5 mL of milliQ water.

9.2.6 3% DMSO in 10 mM HEPES Buffer (10 mL)

1 mL of 100 mM HEPES (Cat. No:, 15630–080, Gibco), 1 mL of 30% DMSO, 1 mL of 20 mM DTT, 1 mL of 0.05% Tween20, 1 mL of 1.5 M KCl was added. pH was adjusted to 7.4, and make up to 10 mL with milliQ water.

9.2.7 2% DMSO in 10 mM HEPES Buffer (1000 mL)

100 mL of 100 mM HEPES (Cat. No:, 15630–080, Gibco), 100 mL of 20% DMSO, 100 mL of 20 mM DTT, 100 mL of 0.05% Tween20, 100 mL of 1.5 M KCl was added. pH was adjusted to 7.4, and make up to 1000 mL with milliQ water.

9.2.8 1% DMSO in 10 mM HEPES Buffer (10 mL)

1 mL of 100 mM HEPES (Cat. No:, 15630–080, Gibco), 1 mL of 10% DMSO, 1 mL of 20 mM DTT, 1 mL of 0.05% Tween20, 1 mL of 1.5 M KCl was added. pH was adjusted to 7.4, and make up to 10 mL with milliQ water.

9.2.9 0% DMSO in 10 mM HEPES Buffer (50 mL)

5 mL of 100 mM HEPES (Cat. No:, 15630–080, Gibco), 5 mL of 20 mM DTT, 5 mL of 0.05% Tween20, 5 mL of 1.5 M KCl was added. pH was adjusted to 7.4, and make up to 50 mL with milliQ water (see **Note 1**).

9.2.10 1.5% DMSO in 10 mM HEPES Buffer (10 mL)

5 mL of HEPES buffer (1% DMSO) and 5 mL of HEPES buffer (2% DMSO) were added to get a final concentration of 1.5%.

9.2.11 2.8% DMSO in 10 mM HEPES Buffer (10 mL)

8 mL of HEPES buffer (3% DMSO) and 2 mL of HEPES buffer (2% DMSO) were added to get a final concentration of 2.8%.

9.2.12 3% Sucrose in 10 mM HEPES Buffer (10 mL)

0.3 g of sucrose (Cat No: 15925, Fisher Scientific) dissolved in 10 mL of HEPES buffer (2% DMSO).

The solution was filtered with a 0.2 micron syringe filter.

9.2.13 Concentrations of Small Molecules Required (Reference)

Small molecules (reference) were prepared at three different concentration, that is, 10 μM, 25 uM, and 50 μM to validate the amine coupling.

9.2.14 Preparation of 10 mM Master Stock

Dry powder (1–2 mg) of small molecule was dissolved in 100% DMSO to make 10 mM. Subsequent substocks were prepared to make working concentration to be tested in SPR (see **Note 2**).

9.2.15 Preparation of 0.5 mM Substock

2 μL from 10 mM stock was added to 38 μL of 100% DMSO.

9.2.16 Preparation of 1.25 mM Substock

5 μL from 10 mM stock was added to 35 μL of 100% DMSO.

9.2.17 Preparation of 2.5 mM Substock

10 μL from 10 mM stock was added to 30 μL of 100% DMSO.

9.2.18 Preparation of 10 uM Working Concentration

15 μL from 0.5 mM substock was added to 735 μL of running buffer (HEPES).

9.2.19 Preparation of 25 uM Working Concentration

15 μL from 1.25 mM substock was added to 735 μL of running buffer (HEPES).

9.2.20 Preparation of 50 uM Working Concentration

15 μL from 2.5 mM substock was added to 735 μL of running buffer (HEPES) (see **Note 3**).

9.2.21 SPR Compatible 96 Well Plate

PioneerFE SPR compatible 96 well plate (Cat. No: P-96-450R-C-S, Axygen).

9.3 Methods

9.3.1 Refractive Index (RI) Transfer

1. DMSO possesses a high refractive index and even low concentrations (<1%) give rise to bulk refractive index responses that dwarf the actual binding signal of small molecules. Consequently, it is necessary to add DMSO to the running buffer before running the experiment to reduce this bulk refractive index offset. Despite best efforts, a difference in sample and assay buffer refractive indices can exist and require solvent calibration to resolve low analyte binding signals (see **Note 4**).
2. To construct a solvent calibration plot, 8 solvent standards (2.80%, 2.61%, 2.43%, 2.24%, 2.05%, 1.87%, 1.67%, and 1.5%) were prepared that encompass the potential solvent concentration range of the sample buffer.
3. These concentrations of DMSO were prepared from 1.5% and 2.8% DMSO combination (Table 9.1).

Table 9.1 Preparation of DMSO concentrations

DMSO concentration	Volume required from 1.5% DMSO solution (μL)	Volume required from 2.8% DMSO solution (μL)
2.80%	345	5
2.61%	300	50
2.43%	250	100
2.24%	200	150
2.05%	150	200
1.87%	100	250
1.67%	50	300
1.5%	5	345

9.3.2 Initiation of SPR Assay

1. The prepared concentration (10 μM, 25 μM, 50 μM) with volume 750 μL each were centrifuged briefly at 12,000 rpm for 10 min to remove insoluble particles (see **Note 5**).
2. After centrifugation, the 700 μL of supernatant was transferred to a new and clean 1.5 mL Eppendorf tube.
3. 200 μL of each concentration is transferred to PioneerFE SPR compatible 96 well plate in triplicates.
4. The plate was sealed with a silicon flap and placed inside the SPR instrument.
5. Program was set up to initiate the assay using Pioneer software (Fig. 9.1).
6. Prior to assay initiation, 4× priming was done with running buffer and MC was initiated.

9.3.3 Micro-calibration (MC)

1. Another internal control parameter that need to be considered was Micro-calibration (MC) (see **Note 6**).
2. We used 2% DMSO in our running buffer during validation based on the prior information from Therma shift assay (TSA).
3. Therefore we used 1% and 3% DMSO in running buffer to generate MC curve (Fig. 9.2) (see **Note 7**).
4. The % error is calculated as follows:

$$\frac{\text{RU at }3\% - \text{RU at }1\%}{\text{Avg RUs of }3\%\text{and }1\%} \times 100$$

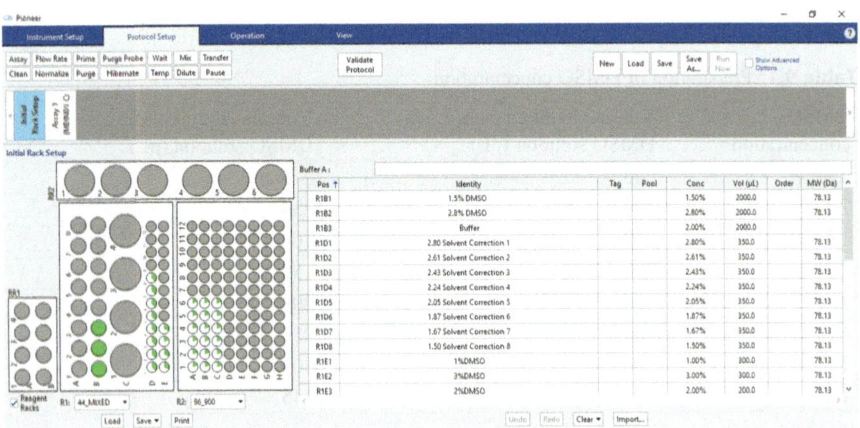

Fig. 9.1 Program setup to initiate SPR assay (Source: Jubilant Biosys)

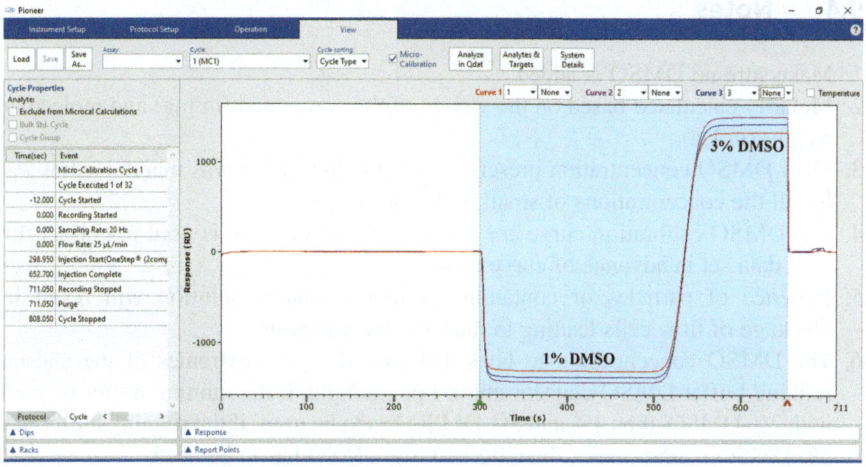

Fig. 9.2 Micro-calibration curve containing 1% and 3% DMSO profiles (Source: Jubilant Biosys)

Table 9.2 Troubleshooting guide for amine coupling of ligand

Problem	Cause	Action
No sensorgram profile	• Cpd precipitation	• Check solubility
	• Air bubble	• Prime with buffer to remove air bubble
Spike observed between steady state and dissociate	• Mismatch between running buffer and RI buffer	• Check running buffer composition, pH, and DMSO. • Prepare fresh running buffer
MC error	• Mismatch between 1%, 3% DMSO buffer, and running buffer	• Prepare fresh buffers

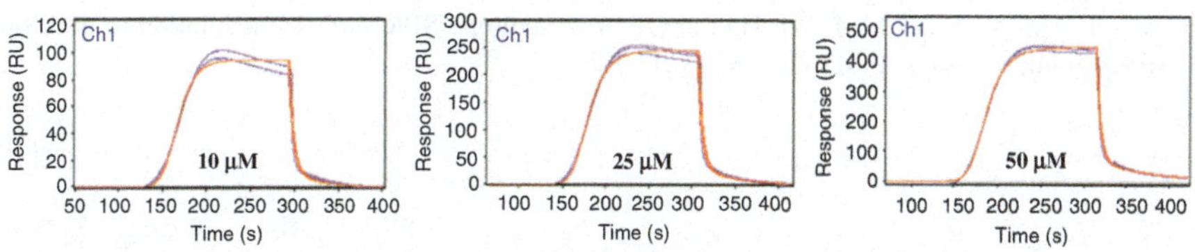

Fig. 9.3 The reference small molecule (reference) tested at 10 µM, 25 µM, 50 µM showed similar K_D values. The RU of these concentration increased with increased tested concentrations (Source: Jubilant Biosys)

5. RI correction was initiated followed by compound testing.
6. The assay was set to be validated when all the three tested concentration showed similar K_D values with increasing RU (Table 9.2 and Fig. 9.3).
7. Detailed analysis explained in Chap. 14.

9.4 Notes

1. Make sure no DMSO is added.
2. Molarity calculated based on the weight of the compound (in mg) and molecular weight (in Da).
3. Final DMSO concentration present in the running buffer was maintained at 2% for all the concentrations of small molecules tested.
4. The DMSO calibration curve construction is required that is used to correct the final data set in advance of curve fitting.
5. Presence of particles or contaminants in the analyte solution will result in blockage of flow cells leading to machine back pressure.
6. The DMSO concentration to be tested should be the extremes of the chosen running buffer DMSO concentration. For example, if the running buffer is fixed with 2% DMSO, then 1% and 3% DMSO has to be used. If the running buffer has 5% DMSO (some cases), then 4% and 6% DMSO has to be used.
7. The error or the difference between the 1% and 3% DMSO buffer profile should be less than 5%.

References

1. Kimple AJ, Muller RE, Siderovski DP et al (2010) A capture coupling method for the covalent immobilization of hexahistidine tagged proteins for surface plasmon resonance. Methods Mol Biol 627:91–100
2. Homaei AA, Sariri R, Vianello F et al (2013) Enzyme immobilization: an update. J Chem Biol 6 (4):185–205
3. Schneider CS, Bhargav AG, Perez JG et al (2015) Surface plasmon resonance as a high throughput method to evaluate specific and non-specific binding of nanotherapeutics. J Control Release 219:331–344
4. Drescher DG, Ramakrishnan NA, Drescher MJ (2009) Surface Plasmon resonance (SPR) analysis of binding interactions of proteins in inner-ear sensory epithelia. Methods Mol Biol 493:323–343
5. Peng L, Damschroder MM, Wu H et al (2014) Bi-epitope SPR surfaces: a solution to develop robust immunoassays. PLoS One 9(11):e112070

Optimization and Validation of Strepavidin/ Biotin-Based Capture of Ligand

10

Krishnakumar Vaithilingam and Sameer Mahmood Zaheer

Abstract

One of the hurdles seen with some proteins is that the analyte binding site has lysine rich residues, which impede biophysical interaction between the ligand and analyte. To overcome this snag, non-covalent capture method was developed that has the same sensitivity as compared to amine coupling. This chapter highlights the affinity capture (non-covalent) capture method used in SPR assay. The most frequent affinity capture technique is streptavidin-biotin capture. Here we provide the protocol that can guide the optimization and validation of streptavidin-biotin capture method using Biocap (specific to PioneerFE SPR).

Keywords

Non-covalent · Streptavidin · Biocap

10.1 Introduction

The BioCap sensor takes advantage of the extremely high affinity of proteins NeutrAvidin™, streptavidin, or avidin for biotin, which has an equilibrium dissociation constant (K_D) of 10^{-12} M [1, 2]. Immobilization of biomolecules by avidin-biotin based methods is very common due to the robustness of the interaction.

The BioCap surface is a carboxylated dextran surface with covalently immobilized NeutrAvidin™. This surface cannot be regenerated without denaturing the immobilized NeutrAvidin™. Regeneration strategies aim to remove the analyte but leave the biotinylated ligand on the surface, making the use of the BioCap

K. Vaithilingam (✉) · S. M. Zaheer
Structural Biology Division, Discovery Biology, Jubilant Biosys Ltd., Bengaluru, Karnataka, India
e-mail: Krishnakumar.V@jubilantbiosys.com; Sameer.Mahmood@jubilantbiosys.com

S. M. Zaheer, R. Gosu (eds.), *Methods for Fragments Screening Using Surface Plasmon Resonance*, https://doi.org/10.1007/978-981-16-1536-8_10

surface for affinity capture more like covalent attachment strategies than other affinity capture approaches. Sensors for biotinylated ligand attachment include SADH chip, which has streptavidin coated to its surface. This binding is highly resistant to heat, pH, and proteolysis that makes it possible to capture a wide range of biotinylated molecules in different environmental conditions. The binding of biotin to streptavidin is reversible in nature. Optimization and validation protocol is presented here with important criteria and troubleshooting with action taken considered at each step.

Several factors influence the effective linking of the ligand to the sensors, as these factors play a pivotal role in the success of the experiment. These factors include temperature of the ligand and machine, buffer pH, and charge mediated concentration of the ligand. This chapter provides a comprehensive detail to validate the ligand immobilized onto the Biocap sensor using a known reference molecule.

10.2 Materials

10.2.1 30% DMSO Solution (100 mL)

30 mL of 100% DMSO (Cat No: 34869, Sigma Aldrich) dissolved in 70 mL of milliQ water.

10.2.2 20% DMSO Solution (100 mL)

20 mL of 100% DMSO (Cat No: 34869, Sigma Aldrich) dissolved in 80 mL of milliQ water.

10.2.3 10% DMSO Solution (100 mL)

10 mL of 100% DMSO (Cat No: 34869, Sigma Aldrich) dissolved in 90 mL of milliQ water.

10.2.4 10 mM DTT (100 mL)

0.15 g of DTT (Cat. No: D9779, Sigma Aldrich) dissolved in 100 mL of milliQ water.

10.2.5 0.05% Tween20 (100 mL)

50μL of 100% Tween20 (Cat. No: P1379, Sigma Aldrich) dissolved in 99.5 mL of milliQ water.

10.2.6 3% DMSO in 10 mM HEPES Buffer (10 mL)

1 mL of 100 mM HEPES (Cat. No: 15630-080, Gibco), 1 mL of 30% DMSO, 1 mL of 20 mM DTT, 1 mL of 0.05% Tween20, 1 mL of 1.5 M KCl was added. pH was adjusted to 7.4, and make up to 10 mL with milliQ water.

10.2.7 2% DMSO in 10 mM HEPES Buffer (1000 mL)

100 mL of 100 mM HEPES (Cat. No: 15630-080, Gibco), 100 mL of 20% DMSO, 100 mL of 20 mM DTT, 100 mL of 0.05% Tween20, 100 mL of 1.5 M KCl was added. pH was adjusted to 7.4, and make up to 1000 mL with milliQ water.

10.2.8 1% DMSO in 10 mM HEPES Buffer (10 mL)

1 mL of 100 mM HEPES (Cat. No: 15630-080, Gibco), 1 mL of 10% DMSO, 1 mL of 20 mM DTT, 1 mL of 0.05% Tween20, 1 mL of 1.5 M KCl was added. pH was adjusted to 7.4, and make up to 10 mL with milliQ water.

10.2.9 0% DMSO in 10 mM HEPES Buffer (50 mL)

5 mL of 100 mM HEPES (Cat. No: 15630-080, Gibco), 5 mL of 20 mM DTT, 5 mL of 0.05% Tween20, 5 mL of 1.5 M KCl was added. pH was adjusted to 7.4, and make up to 50 mL with milliQ water (see **Note 1**).

10.2.10 1.5% DMSO in 10 mM HEPES Buffer (10 mL)

5 mL of HEPES buffer (1% DMSO) and 5 mL of HEPES buffer (2% DMSO) were added to get a final concentration of 1.5%.

10.2.11 2.8% DMSO in 10 mM HEPES Buffer (10 mL)

8 mL of HEPES buffer (3% DMSO) and 2 mL of HEPES buffer (2% DMSO) were added to get a final concentration of 2.8%.

10.2.12 3% Sucrose in 10 mM HEPES Buffer (10 mL)

0.3 g of sucrose Cat No: in 10 mL of HEPES buffer (2% DMSO).

The solution was filtered with a 0.22μm syringe filter.

10.2.13 Concentrations of Small Molecules Required (Reference)

Small molecules (reference) were prepared at three different concentration, that is, 10μM, 25μM, and 50μM to validate the biotin-strepavidin capture.

10.2.14 Preparation of 10 mM Master Stock

Dry powder (1–2 mg) of small molecule was dissolved in 100% DMSO to make 10 mM. Subsequent substocks were prepared to make working concentration to be tested in SPR (see **Note 2**).

10.2.15 Preparation of 0.5 mM Substock

2μL from 10 mM stock was added to 38μL of 100% DMSO.

10.2.16 Preparation of 1.25 mM Substock

5μL from 10 mM stock was added to 35μL of 100% DMSO.

10.2.17 Preparation of 2.5 mM Substock

10μL from 10 mM stock was added to 30μL of 100% DMSO.

10.2.18 Preparation of 10 uM Working Concentration

Preparation of 10μM working concentration: 15μL from 0.5 mM substock was added to 735μL of running buffer (HEPES).

10.2.19 Preparation of 25 uM Working Concentration

15μL from 1.25 mM substock was added to 735μL of running buffer (HEPES).

10.2.20 Preparation of 50 uM Working Concentration

15μL from 2.5 mM substock was added to 735μL of running buffer (HEPES) (see **Note 3**).

10.2.21 SPR Compatible 96 Well Plate

PioneerFE SPR compatible 96 well plate (Cat. No: P-96-450R-C-S, Axygen).

10.3 Methods

10.3.1 Refractive Index (RI) Correction

1. DMSO possesses a high refractive index and even low concentrations (<1%) give rise to bulk refractive index responses that dwarf the actual binding signal of small molecules. Consequently, it is necessary to add DMSO to the running buffer before running the experiment to reduce this bulk refractive index offset. Despite best efforts, a difference in sample and assay buffer refractive indices can exist and require solvent calibration to resolve low analyte binding signals (see **Note 4**).
2. To construct a solvent calibration plot, 8 solvent standards (2.80%, 2.61%, 2.43%, 2.24%, 2.05%, 1.87%, 1.67%, and 1.5%) were prepared that encompass the potential solvent concentration range of the sample buffer.
3. These concentrations of DMSO was prepared from 1.5% and 2.8% DMSO combination (Table 10.1).

10.3.2 Initiation of SPR Assay

1. The prepared concentration (10μM, 25μM, 50μM) with volume 750μL each were centrifuged briefly at 12000 rpm for 10 min to remove particles (see **Note 5**).
2. After centrifugation, the supernatant was transferred to a new and clean 1.5 mL Eppendorf tube.
3. 200μL of each concentration is transferred to PioneerFE SPR compatible 96 well plate in triplicates.
4. The plate is sealed with a silicon flap and placed inside the SPR instrument.

Table 10.1 Preparation of DMSO concentrations

DMSO concentration	Volume required from 1.5% DMSO solution (μL)	Volume required from 2.8% DMSO solution (μL)
2.80%	345	5
2.61%	300	50
2.43%	250	100
2.24%	200	150
2.05%	150	200
1.87%	100	250
1.67%	50	300
1.5%	5	345

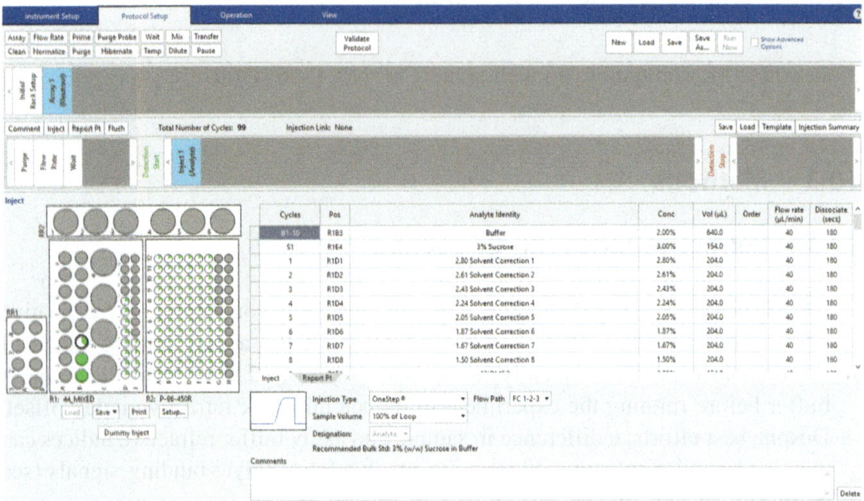

Fig. 10.1 Program setup to initiate SPR assay (Source: Jubilant Biosys)

5. Program was set up to initiate the assay using Pioneer software (Fig. 10.1).
6. Prior to assay initiation, 4× priming was done with running buffer and MC was initiated.

10.3.3 Micro-calibration (MC)

1. Another internal control parameter that need to be considered is Micro-calibration (MC) (see **Note 6**).
2. We used 2% DMSO in our running buffer during validation based on the prior information from Therma shift assay (TSA).
3. Therefore we used 1% and 3% DMSO in running buffer to generate MC curve (Fig. 10.2) (see **Note 7**).
4. The % error is calculated as follows:

$$\frac{\text{RU at }3\% - \text{RU at }1\%}{\text{Avg RUs of }3\%\text{and }1\%} \times 100$$

5. RI correction was initiated followed by compound testing.
6. The assay was set to be validated when all the three tested concentration showed similar K_D values with increasing RU (Table 10.2 and Fig. 10.3).
7. Refer to Chap. 14 for detailed analysis.

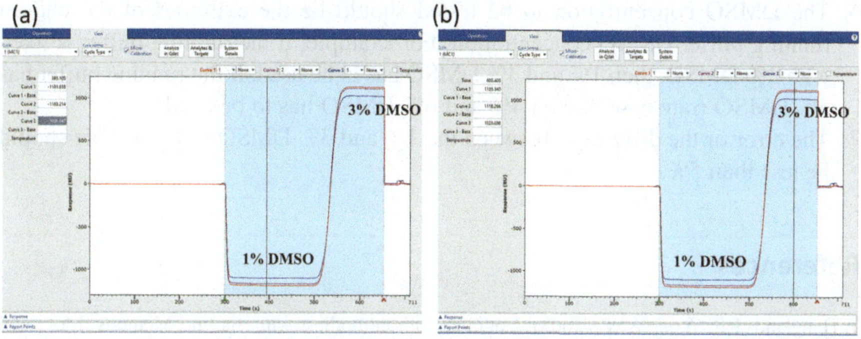

Fig. 10.2 Micro-calibration Curve containing 1% and 3% DMSO profiles. Crosshair positioned at 1% DMSO reading and 3% DMSO reading shown in (**a**) & (**b**) respectively (Source: Jubilant Biosys)

Table 10.2 Troubleshooting guide for biotin-streptavidin capture

Problem	Cause	Action
Low RUs	• High flow rate	• Reduce the flow rate
	• Low protein conc.	• Increase protein conc.
	• Running buffer pH and protein pI mismatch	• pH should be closer to the pI of the protein

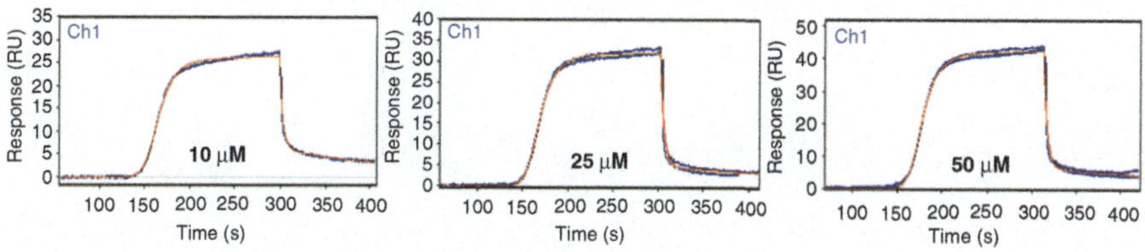

Fig. 10.3 The small molecule (reference) tested at 10 μM, 25 μM, 50 μM showed similar KD values. The RU of these concentration increased with increased tested concentrations

10.4 Notes

1. Make sure no DMSO is added.
2. Molarity is calculated based on the weight of the compound (in mg) and molecular weight (in Da).
3. Final DMSO concentration present in the running buffer was maintained at 2% for all the concentrations of small molecules tested.
4. The DMSO calibration curve construction is required that is used to correct the final data set in advance of curve fitting.
5. Presence of particles or contaminants in the analyte solution will result in blockage of flow cells.

6. The DMSO concentration to be tested should be the extremes of the chosen running buffer DMSO concentration. For example, if the running buffer is fixed with 2% DMSO, then 1% and 3% DMSO has to be used. If the running buffer has 5% DMSO (rare cases), then 4% and 6% DMSO has to be used.
7. The error or the difference between the 1% and 3% DMSO buffer profile should be less than 5%.

References

1. Haugland RP, You WW (2008) Coupling of antibodies with biotin. Methods Mol Biol 418:13–24
2. Orelma H, Johansson L-s, Filpponen I et al (2012) Generic method for attaching biomolecules via avidin-biotin complexes immobilized on films of regenerated and nanofibrillar cellulose. Biomacromolecules 13(9):2802–2810

Preparation of Fragments for Screening in SPR

11

Sameer Mahmood Zaheer and Aswathy Pillai

Abstract

Sartorius (formerly ForteBio) SPR system is highly sensitive to measure the binding kinetics of small molecule fragments. Important factors responsible for successfully screening small molecule fragments lie with the proper sample preparation before injecting into the SPR system. One of the major limitations of the HTS campaign using SPR with small molecules is solubility. Fragments are more soluble than small molecules. However, if even one or a small fraction of fragments in a screening plate are less soluble, it can lead to blockage of SPR system capillaries. This chapter provides optimal details required for the preparation of fragments to minimize clogging.

Keywords

HTS campaign · Solubility · Fragments

11.1 Introduction

Screening of small molecule fragments through the HTS campaign has helped in the improvisation of hit identification in the drug discovery program [1]. Fragments screening with SPR provides an added advantage whereby the characterization of these fragments can be identified based on their ligand efficiencies [2]. Preparation of these fragments is a very critical step during the screening process as these fragments are the building blocks for developing a small molecule. The factors that need to be considered are concentration of fragments to be tested in SPR, and

S. M. Zaheer (✉) · A. Pillai
Structural Biology Division, Discovery Biology, Jubilant Biosys Ltd., Bengaluru, Karnataka, India
e-mail: Sameer.Mahmood@jubilantbiosys.com

S. M. Zaheer, R. Gosu (eds.), *Methods for Fragments Screening Using Surface Plasmon Resonance*, https://doi.org/10.1007/978-981-16-1536-8_11

the amount of DMSO required to avoid precipitation. Here we provide the best possible methods to prepare fragments to be used in SPR.

11.2 Materials

1. 96-well SPR compatible assay plate (Cat No: P-96-450R-C-S, Axygen).
2. 384-well SPR compatible assay plate (Cat No: P-384-240SQ-C-S, Axygen).
3. 96-well storage plate (Cat No: 3368, Costar).
4. 96-well deep (1.1 mL capacity) plate (Cat No: P-DW-11-C, Axygen).
5. Aluminium foil lids (Cat No: 538619, Beckman Coulter).
6. Multichannel pipette (10 μL) (8-channel).
7. Multichannel pipette (100 μL) (8-channel).
8. Multichannel pipette (300 μL) (8-channel).
9. Sample addition reservoir—50 mL.

11.2.1 30% DMSO Solution (100 mL)

30 mL of 100% DMSO (Cat No: 34869, Sigma Aldrich) dissolved in 70 mL of milliQ water.

11.2.2 20% DMSO Solution (100 mL)

20 mL of 100% DMSO (Cat No: 34869, Sigma Aldrich) dissolved in 80 mL of milliQ water.

11.2.3 10% DMSO Solution (100 mL)

10 mL of 100% DMSO (Cat No: 34869, Sigma Aldrich) dissolved in 90 mL of milliQ water.

11.2.4 10 mM DTT (100 mL)

0.15 g of DTT (Cat. No: D9779, Sigma Aldrich) dissolved in 100 mL of milliQ water.

11.2.5 0.05% Tween20 (100 mL)

50 μL of 100% Tween20 (Cat. No: P1379, Sigma Aldrich) dissolved in 99.5 mL of milliQ water.

11.2.6 3% DMSO in 10 mM HEPES Buffer (10 mL)

1 mL of 100 mM HEPES (Cat. No:, 15630–080, Gibco), 1 mL of 30% DMSO, 1 mL of 20 mM DTT, 1 mL of 0.05% Tween20, 1 mL of 1.5 M KCl was added. pH was adjusted to 7.4, and make up to 10 mL with milliQ water.

11.2.7 2% DMSO in 10 mM HEPES Buffer (1000 mL)

100 mL of 100 mM HEPES (Cat. No:, 15630–080, Gibco), 100 mL of 20% DMSO, 100 mL of 20 mM DTT, 100 mL of 0.05% Tween20, 100 mL of 1.5 M KCl was added. pH was adjusted to 7.4, and make up to 1000 mL with milliQ water.

11.2.8 1% DMSO in 10 mM HEPES Buffer (10 mL)

1 mL of 100 mM HEPES (Cat. No:, 15630–080, Gibco), 1 mL of 10% DMSO, 1 mL of 20 mM DTT, 1 mL of 0.05% Tween20, 1 mL of 1.5 M KCl was added. pH was adjusted to 7.4, and make up to 10 mL with milliQ water.

11.2.9 0% DMSO in 10 mM HEPES Buffer (50 mL)

10 mL of 100 mM HEPES (Cat. No:, 15630–080, Gibco), 10 mL of 20 mM DTT, 10 mL of 0.05% Tween20, 10 mL of 1.5 M KCl was added. pH was adjusted to 7.4, and make up to 100 mL with milliQ water. (see **Note 1**).

11.2.10 1.5% DMSO in 10 mM HEPES Buffer (10 mL)

5 mL of HEPES buffer (1% DMSO) and 5 mL of HEPES buffer (2% DMSO) were added to get a final concentration of 1.5%.

11.2.11 2.8% DMSO in 10 mM HEPES Buffer (10 mL)

8 mL of HEPES buffer (3% DMSO) and 2 mL of HEPES buffer (2% DMSO) were added to get a final concentration of 2.8%.

11.2.12 3% Sucrose in 10 mM HEPES Buffer (10 mL)

0.3 g of sucrose (Cat No: 15925, Fisher Scientific) dissolved in 10 mL of HEPES buffer (2% DMSO).

The solution was filtered with a 0.22 μm syringe filter.

11.3 Methods

11.3.1 Fragment Master Plate (10 mM)

1. Fragments were received from the vendor in a 96 well format and stored at −20 °C. (see **Note 2**).
2. Concentration of received fragments were 10 mM stock in 100 μL (100% DMSO). (see **Note 3**).
3. 80 fragments were distributed from A2 to H11 wells in a 96-well plate. (see **Note 4**)
4. On the day of initiation of screening, master plate (10 mM stock) was removed from −20 °C freezer thawed at room temperature for 1 hour or until the fragments stock is completely thawed.
5. Once thawed, the plate is briefly spun at 2500 rpm for 10 min at room temperature.
6. Subsequent substock were prepared: 5 mM and 100 μM.

11.3.2 Substock Plate (5 mM)

1. 30 μL of 10 mM stock was added to 30 μL of 100% DMSO to make a 5 mM substock solution. (Fig. 11.1b)
2. 60 μL of the solution is mixed to get a homogenous solution.

11.3.3 SPR Assay Plate (100 μM)

1. 10 μL of 5 mM substock was added SPR assay plate containing 490 μL of running buffer without DMSO (Fig.11.1c). (see **Note 5**).

11.3.4 Precipitation Check

1. Another assay plate is prepared in a UV compatible plate to measure absorbance at 590 nm.
2. All absorbance readings were compared to positive control (100% DMSO).
3. The precipitated fragment wells were excluded from screening (Table 11.1).

11.4 Notes

1. Make sure no DMSO is added.
2. Care should be taken, the plates are not thawed until screening.
3. Substock and assay plates were generated from this master plate.

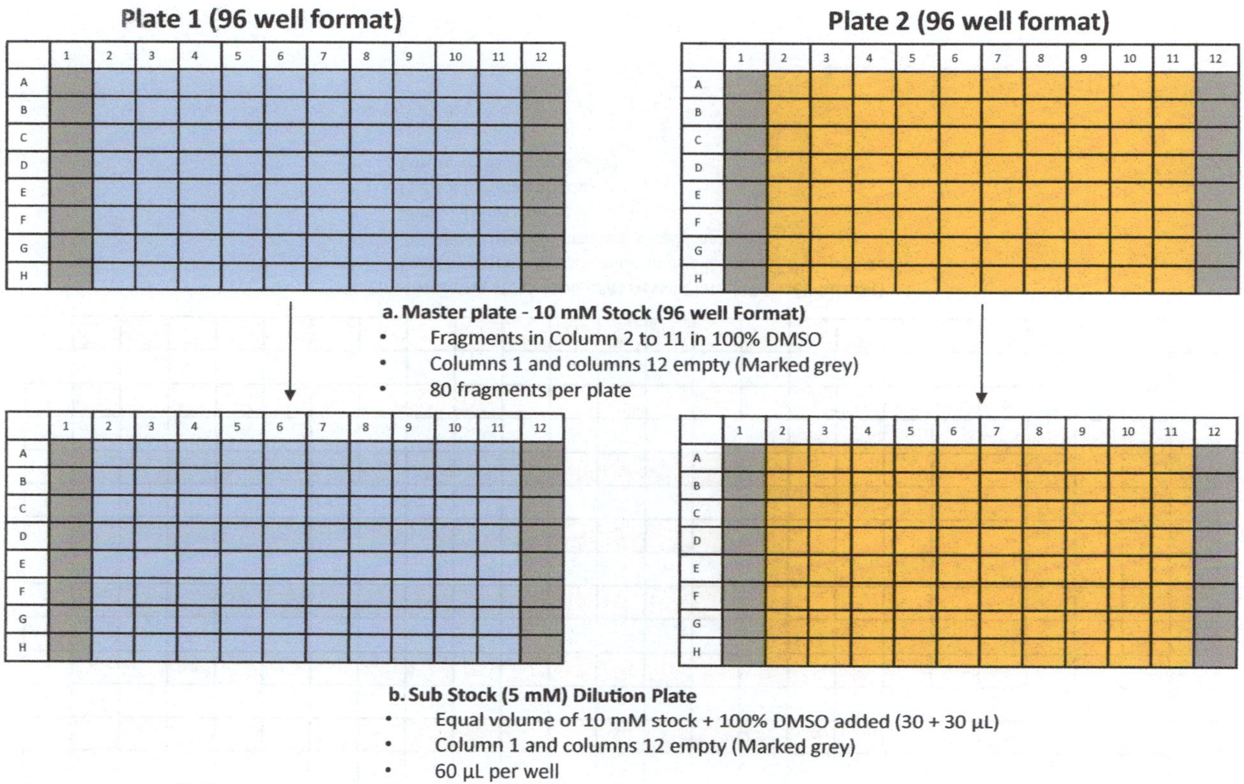

Fig. 11.1 Preparation of Fragments for SPR. (**a**) Master plate—10 mM Stock (96 well format). (1) Fragments in Columns 2–11 in 100% DMSO. (2) Columns 1 and 12 empty (marked grey). (3) 80 fragments per plate. (**b**) Sub Stock (5 mM) dilution plate. (1) Equal volume of 10 mM stock +100% DMSO added (30 + 30 µL). (2) Columns 1 and 12 empty (marked grey). (3) 60 µL per well. (**c**) Final working stock (100 µM) Assay Plate (384 well format). (1) Each fragment was stamped in duplicates (150 µL per well). (2) 2% DMSO was maintained for all fragments. (3) Plate 1 (blue shade). (4) Plate 2 (orange shade). (5) Empty wells (grey shade) (Source: Jubilant Biosys)

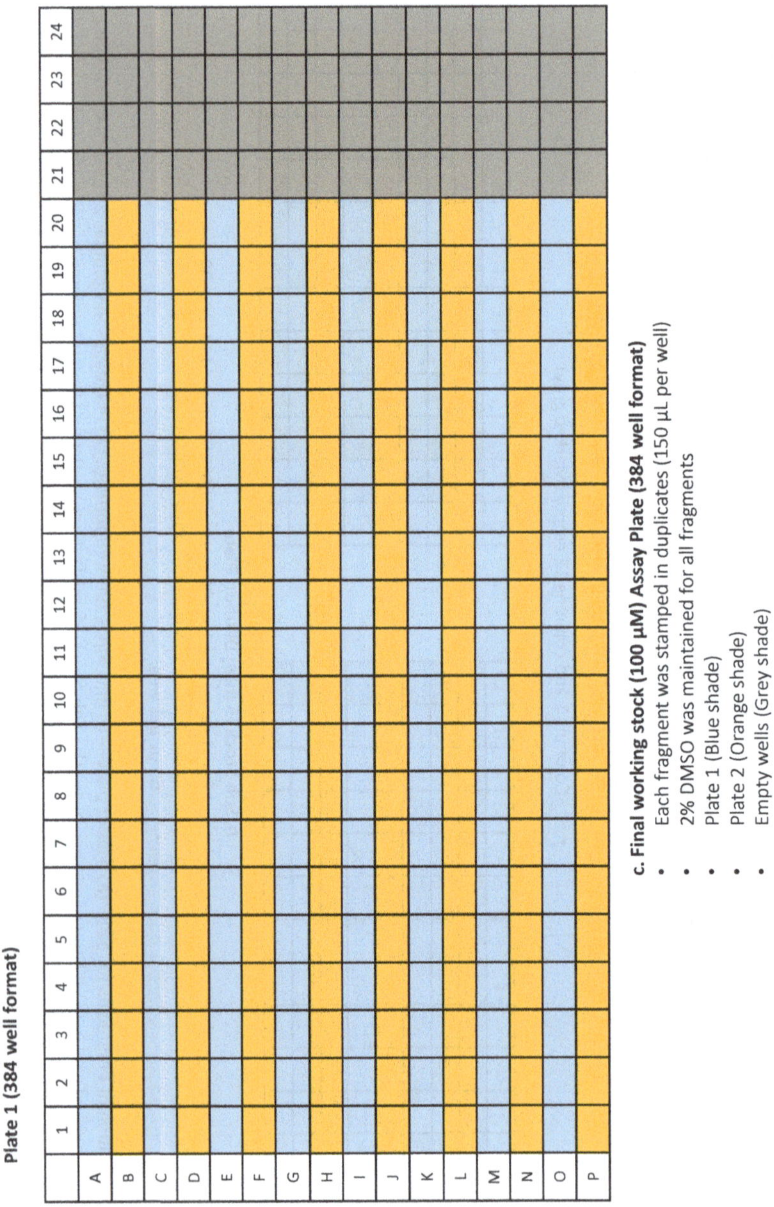

Fig. 11.1 (continued)

Table 11.1 Troubleshooting guide for Preparation of Fragment for Screening

Problem	Cause	Action
Master plate identification	Label missing	Confirm with vendor's master list
Master plate sample leakage	Loose caps	Tighten it properly after thawing
Master plate low volume	Sample in caps	Spin the plates after thawing
Assay plate solution do not settle down	Air bubble	Spin the plate after dispensing the solution in assay
More volume (>150 μL) in the well compared to other wells (150 μL)	Sample gets transferred from an adjacent well	Use narrow tips and dispense deep inside the assay plate well

4. Columns 1 and 12 are empty.
5. The final concentration of DMSO in the running buffer was kept at 2%.

References

1. Larsen MJ, Larsen SD, Fribley A et al (2014) The role of HTS in drug discovery at the University of Michigan. Comb Chem High Throughput Screen 17(3):210–230
2. Schultes S, de Graaf C, Haaksma EEJ et al (2011) Ligand efficiency as a guide in fragment hit selection and optimization. Drug Discov Today Technol 7(3):e147–e202

Screening of Fragments in SPR

12

Sameer Mahmood Zaheer and Aswathy Pillai

Abstract

High throughput screening using the SPR system has revolutionized drug discovery in identifying hits. Several SPR platforms are available with their respective software to setup the screening cascade. We used Pioneer software to input various parameters required for a successful screening.

Keywords

Plate map · Fragments

12.1 Introduction

The detection and characterization of fragment binding are dependent on events related to setup of the procedures. This is one of the critical steps in process of identifying fragment hits as a minor error leads to a change in the identity of the fragment list. This chapter provides the setting up of protocols needed for the screening of fragments. Plate map is designed to transfer the fragments to appropriate wells. The volume of the analyte and concentration of ligand are fed into the software. The reference channel is selected for background correction, which will be used later for analysis.

S. M. Zaheer (✉) · A. Pillai
Structural Biology Division, Discovery Biology, Jubilant Biosys Ltd., Bengaluru, Karnataka, India
e-mail: Sameer.Mahmood@jubilantbiosys.com

S. M. Zaheer, R. Gosu (eds.), *Methods for Fragments Screening Using Surface Plasmon Resonance*, https://doi.org/10.1007/978-981-16-1536-8_12

12.2 Materials

Pioneer software version 4.3.1 used for setting up and running assays.

12.3 Methods

Refractive Index (RI) correction and Micro-calibration (MC) are important components in Pioneer based SPR systems.

RI and MC have been explained in (Sects. 9.3.1 and 9.3.2.1 respectively).

12.3.1 Setup of Protocol for Screening

1. The Protocol Setup page in Pioneer software allows the user to view, modify, and write protocols. The Pioneer software enables the user to construct even the most complex protocols quickly (Fig. 12.1).
2. The Protocol Setup environment consists of a toolbar where the system commands can be selected and then dropped into a protocol time strip. (see Note 1).
3. The number of wells was selected—we used 320 wells (see Note 2) (Fig. 12.1).
4. The wells were defined with the fragment identity, the concentration of fragments (100 μM), volume per well (150 μL), and molecular weight.
5. In the general tab, the details of assay name, cycle order (sequential), reference channel (Ch2), and sampling rate (20 Hz) were provided (Fig. 12.2).
6. Assay setup tab was selected and provided information related to flowrate (150 μL/min), Dissociation time (120 s).

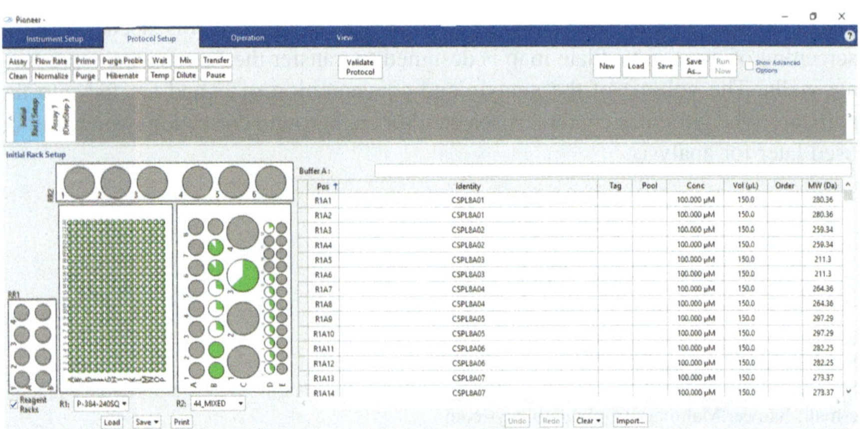

Fig. 12.1 SPR protocol setup page (Source: Jubilant Biosys)

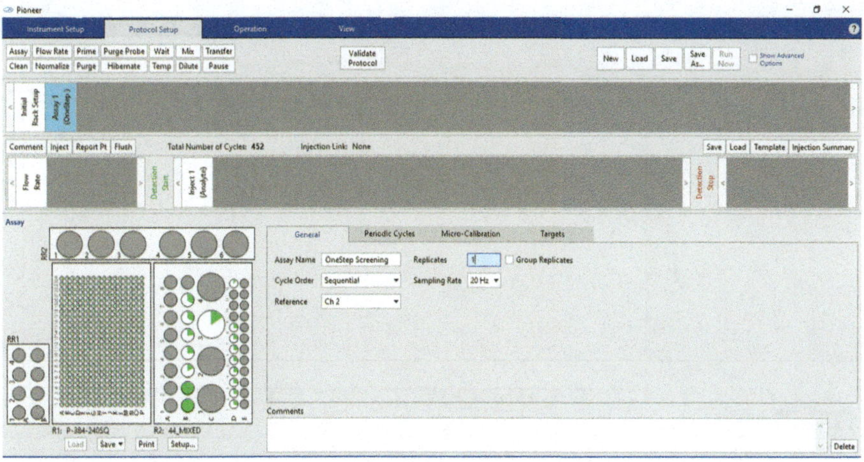

Fig. 12.2 Details for an SPR screening protocol with reference channel and sampling rate (Source: Jubilant Biosys)

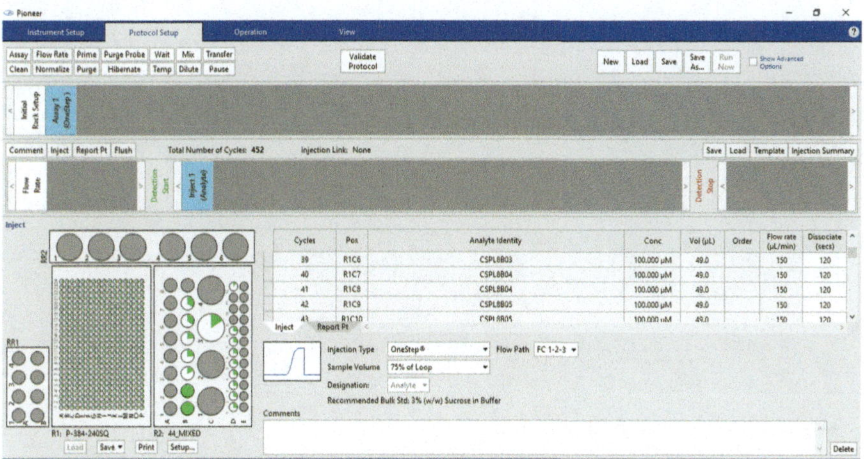

Fig. 12.3 Details for an SPR screening protocol with flow path set (Source: Jubilant Biosys)

7. Sample volume was set at 75% of the loop (see Note 3) with OneStep injected selected. The flowpath is set at FC-1-2-3 (Fig. 12.3).
8. The protocol was saved (pioneer protocol file) and imported to run the assay.

12.3.2 Initiation of Screening

1. In the operation tab, the "Run" was selected where the saved pioneer protocol file was imported (Fig. 12.4).

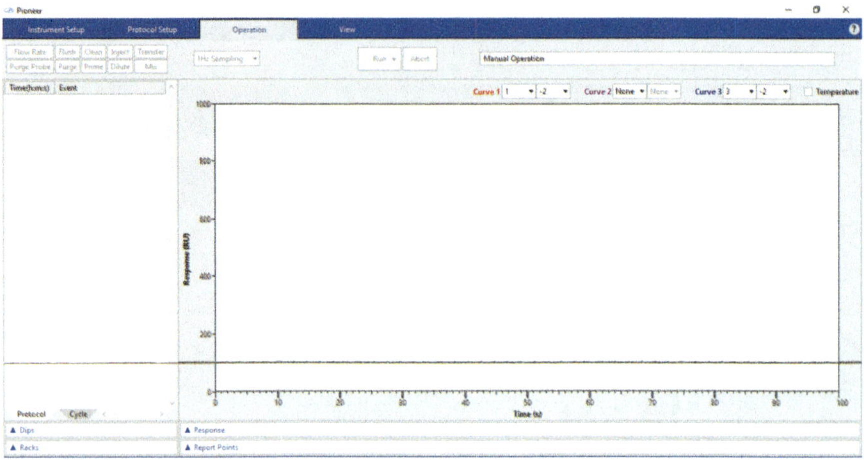

2. The screening was initiated, the assay was completed in ~30 h.
3. Once complete the screening file was saved as "SPR data file" (Fig. 12.5 and Table 12.1).

12.4 Notes

1. Protocols can be constructed, checked for errors, saved, edited, and reloaded.
2. Eighty fragments were present in a 96-well master plate—explained in Chap. 11.
3. The amount of fragment volume assigned.

Table 12.1 Troubleshooting guide for screening of fragments

Problem	Cause	Action
Assay does not initiate run	• Door could be open	• Close the door
	• Other SPR file could be open	• Close all SPR files
Instead of green color, black color displayed in wells chosen for screening	• Inadequate sample volume per well inserted while programming	• Check the volume column
	• Fragment identity is given twice	• Correct the fragment identity
	• Injection loop set to 100% (default)	• Change the injection loop to 75% • Lower the loop percent, lower the sample volume injected

Identification of Fragment Hits

13

Sameer Mahmood Zaheer and Ramachandraiah Gosu

Abstract

This chapter focuses on analysis of SPR data, which was generated from fragments screening. We used "Qdat" analysis software to process the binding response curves for kinetic and affinity analysis. The user-friendly interface guides the user through a series of data transformations that produce data sets that are then fitted with standard kinetic or affinity models. A key feature is the ability to rapidly perform single or double referencing of the data set in advance of model fitting. A tool that enables correction of the excluded volume effect when using solvents such as DMSO is also incorporated. The models include the simple dissociation rate model, the pseudo-first-order kinetic model for a 1:1 interaction, the two compartment model (mass transport limitation), bivalent analyte model and two-state model Here we use point studies module in Qdat software to identify hits.

Keywords

Qdat analysis tool · Pseudo-first-order kinetic model · Bivalent analyte model · Two-state model · Gradient injection · Box-Whisker statistics · Ligand efficiency · Heavy atoms

S. M. Zaheer (✉) · R. Gosu
Structural Biology Division, Discovery Biology, Jubilant Biosys Ltd., Bengaluru, Karnataka, India
e-mail: Sameer.Mahmood@jubilantbiosys.com; ramachandraiah.gosu@jubilanttx.com

S. M. Zaheer, R. Gosu (eds.), *Methods for Fragments Screening Using Surface Plasmon Resonance*, https://doi.org/10.1007/978-981-16-1536-8_13

13.1 Introduction

Hit identification from fragments is an important feature in the process of SPR screening. Fragments have low molecular weight and have the propensity of weak binding affinity towards the ligand [1]. These fragments need to be differentiated between true versus false positive hits. This places constrains on analysis of the data.

The Hit selection software (Qdat) feature is enabled for PioneerFE (fragment edition) instrument models. The purpose of the hit selection feature is to enable rapid and accurate identification of fragment hits in the setting of a primary screen assay. The necessary elements for Hit selection are: injections of fragments at sufficient concentration to elicit response (typically 100–500 μM), a positive control analyte at a near-saturating concentration injected periodically, a negative control analyte (optional) injected periodically, at least 96 fragment analytes and preferably 384 or more. The hit selection processes primary SPR screening data by applying a LOESS correction to the positive control and response data to account for drift, activity decay, aberrations, and so on. The Qdat analysis software uses "point studies" with "Box-Whisker" statistical tool to generate hits. The fundamental principle of using Box-Whisker, is finding an outlier in a distribution by computing IQR (inter quartile range).

13.2 Materials

1. Qdat analysis software version (4.3.1 build 2)—Data analysis tool (Pall Fortebio 2017).
2. Graphpad Prism Version 5.5.

13.3 Methods

1. SPR data were imported in Qdat analysis software.
2. This user-friendly software identifies hits based on an in-build statistical tool.
3. Point studies (used for HTS) were clicked and from the drop-down menu "HIT selection" was selected.
4. Positive and negative controls were confirmed, followed by checking in "Box-Whisker" statistical selection method. (Fig. 13.1). (see **Note 1**).
5. Once fragment hits were identified, the fragment's heavy atoms (non-hydrogen atoms) were counted and input into the Qdat software.
6. Qdat generates the ligand efficiency (LE) with K_D. (see **Note 2**).
7. Hits were plotted with LE versus K_D using graphpad prism software and were utilized to array the fragments. (Fig. 13.2) (see **Note 3**).
8. These fragment "Hits" were further subjected to full kinetic binding analysis (Table 13.1).

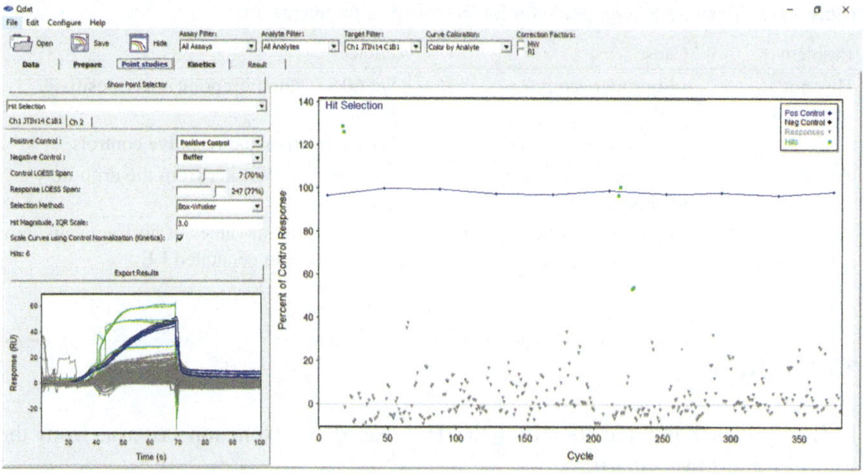

Fig. 13.1 Identification of hits using Qdat analysis software (Source: Jubilant Biosys)

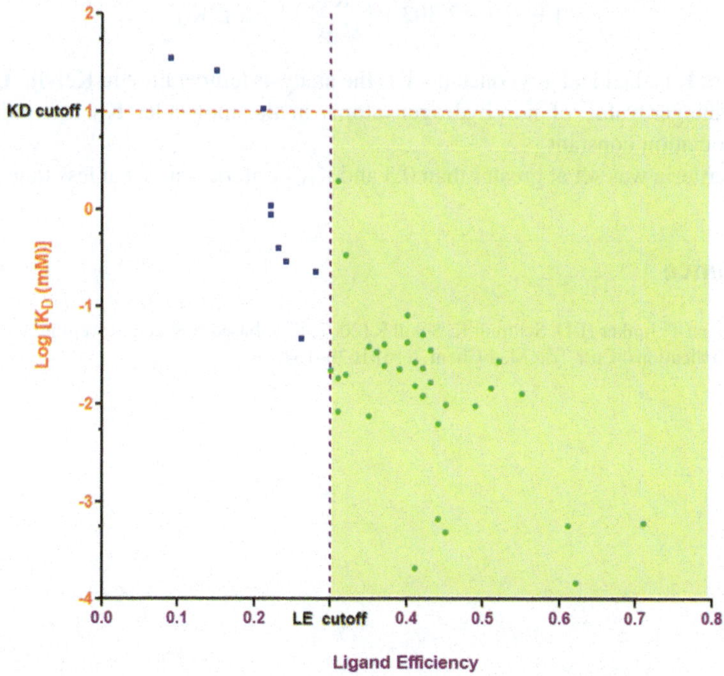

Fig. 13.2 Hits identified and plotted against ligand efficiency vs K_D values from Qdat software. The scatter plot graph was generated using graphpad prism software (Source: Jubilant Biosys)

Table 13.1 Troubleshooting guide for identification of fragments hits

Problem	Cause	Action
Hits not identified	• Improper setup for analysis	• Set 60 s (saturation point) over positive control
	• Controls not selected	• Check positive and negative controls
Several hits identified	• Combined selection method selected	• Choose Box-Whisker from the drop-down menu
Hits identified without LE	• Non-input of heavy atom counts of fragments	• Add heavy atom values of hit fragments into Qdat software to generated LE

13.4 Notes

1. Box-Whisker method identifies the Hits based on exclusion criteria, from the upper and lower quartiles.
2. Formula for calculating ligand efficiency:

$$\mathrm{LE} = \left(-2.303 * \left(\frac{\mathrm{RT}}{\mathrm{HAC}}\right)\right) * \log(\mathrm{K_D})$$

   where **R** is the ideal gas constant, **T** is the analysis temperature in Kelvin, **HAC** is the integer value of non-hydrogen atoms in the molecule, $\mathbf{K_D}$ is equilibrium dissociation constant
3. LE criteria was set at greater than 0.3 and K_D's criteria was set at less than 1 mM.

Reference

1. Neumann T, Junker H-D, Schmidt K, Sekul R (2007) SPR-based fragment screening: advantages and applications. Curr Top Med Chem 7(16):1630–1642

Data Analysis and Confirmation of Hits

14

Sameer Mahmood Zaheer and Ramachandraiah Gosu

Abstract

Fragments generated from SPR HTS campaign were subjected to full binding kinetic analysis. This is the second set of filters placed to eliminate false positive hits, if any, been selected in HTS. Binding parameters like kon, koff, and K_D are highly essential for developing a small molecule. This full kinetic analysis uses Qdat software with the Two-State model as compared to HTS. This is one of the critical stages of drug discovery primary screening using SPR. An error at this point in identifying a true hit will have a significant impact on the generation of starting material and mislead medicinal chemists. This final chapter provides in detail, the generation of kinetic parameters of these hits.

Keywords

Two-state model · Dapp (apparent diffusion co-efficient) · Nagg (aggregation number) · QDat analysis · Kinetic map

14.1 Introduction

Quantitative assessment of quality fragment hits is critical in the process of the development of small molecules. This assessment not only helps in generation of novel small molecule but also drive the structure activity relationships (SAR) for first-in-class drug targets.

In OneStep data, the concentration of analyte changes continuously during the injection, and versions of the simple model and the complex models have been

S. M. Zaheer (✉) · R. Gosu
Structural Biology Division, Discovery Biology, Jubilant Biosys Ltd., Bengaluru, Karnataka, India
e-mail: Sameer.Mahmood@jubilantbiosys.com; ramachandraiah.gosu@jubilanttx.com

S. M. Zaheer, R. Gosu (eds.), *Methods for Fragments Screening Using Surface Plasmon Resonance*, https://doi.org/10.1007/978-981-16-1536-8_14

adapted accordingly. For all reversible fragments, the most suitable model to fit data would be the Two-state model, which represents upon analyte binds to ligand forms reversible complex, which goes into a different state of reversible complex when the analyte is dissociated. The analyte diffusion coefficient is a critical component of the analysis of OneStep® data because it defines the shape of the concentration gradient. There are two new factors, Dapp (Apparent Diffusion Coefficient) and Nagg (Aggregation Number). The Dapp defaults to the value determined from the analyte molecular weight. In this case, Nagg will be at 1, indicating the analyte is a monomer of its ideal MW. Both of these can be fitted by selecting the Dapp column and floating these values.

14.2 Materials

1. Pioneer and Qdat analysis software version (4.3.1 build 2)—Data analysis tool (Pall Fortebio 2017) was used for set up, run assay, and analysis.

14.3 Methods

1. Refractive Index (RI) correction and Micro-calibration (MC) are important components in Pioneer-based SPR system.
2. RI and MC have been explained in Sects. 9.3.1 and 9.3.2.1 respectively).

14.3.1 Setup of Protocol for Screening

1. The Protocol Setup page is explained in Sect. 12.3.1.
2. 100μM fragments were tested for full kinetics. (see **Note 1**).

14.3.2 Initiation of Screening

1. In the operation tab, the "Run" was selected, and the saved pioneer SPR protocol file (.protocol) was imported.
2. The screening was initiated and assay was completed in ~30 h.
3. The file was saved as "SPR data file" (.spr) (Fig. 14.1).

14.3.3 Data Analysis

1. "SPR data file" was opened using "Analyze on Qdat" tab (see **Note 2**).
2. Data tab (Fig. 14.2) was selected (see **Note 3**).
3. Color coding of the data was adjusted by changing the selection of the Curve Coloration drop-down box (see **Note 4**).

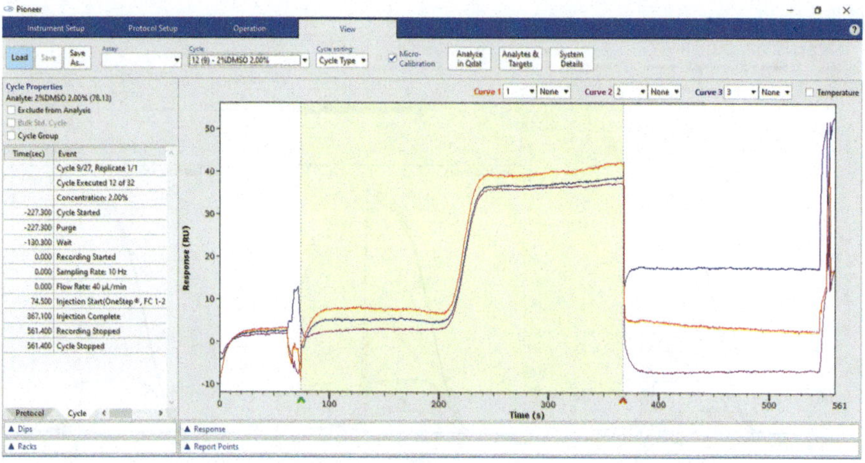

Fig. 14.1 SPR data file (Source: Jubilant Biosys)

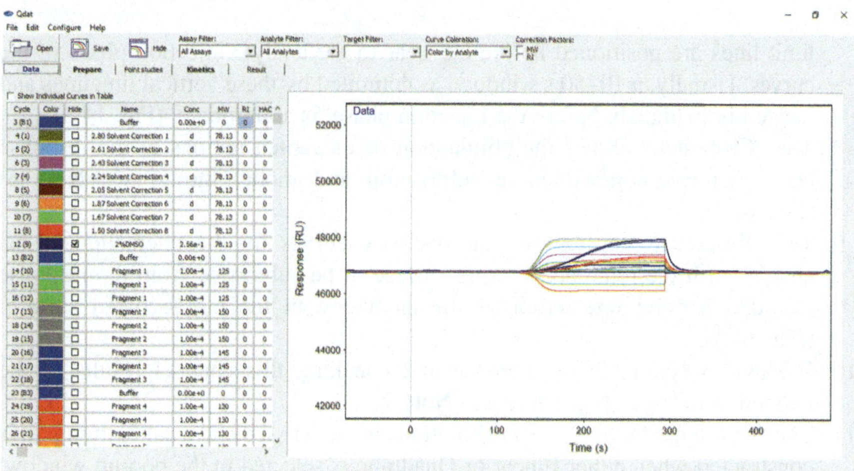

Fig. 14.2 Qdat analysis file—data tab (Source: Jubilant Biosys)

4. Data opened in Qdat were automatically Despiked, meaning that an algorithm scans the data and removes/smooths a window of data around perceived spikes (see **Note 5**).
5. The concentrations of the DMSO standard samples were set to value "d" to designate them as solvent standards.
6. The "Prepare tab" contains all the components for data referencing, aligning, and calibrating. These steps are from left to right: Zero, Crop, Align, Reference, RI Calibration, and Blanks. (Fig. 14.3).
7. The "Zero step" allows the set of response curves to be y-normalized so that a zero baseline is set for all curves in the set. The left (green) and right (red) time

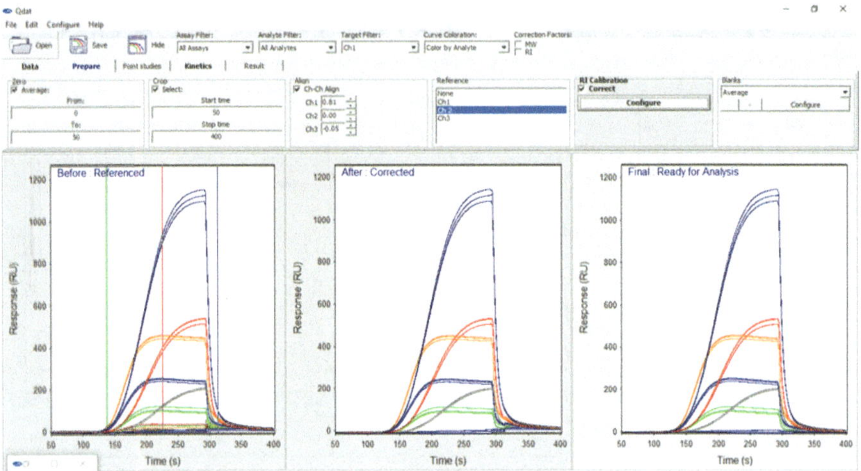

Fig. 14.3 Qdat analysis—prepare section with zero, crop, reference tabs (Source: Jubilant Biosys)

limit lines are positioned before but near to the analyte injection phase of the curves. Usually, a 10–50 s window, as delimited by these vertical timelines and located immediately before the injection phase, is appropriate. (Fig. 14.3).

8. The "Crop step" allows the elimination of unwanted regions of the response curve before reference curve subtraction and model fitting (see **Note 6**) (Fig. 14.3).
9. The "Reference step" allows the response curves for the interaction of the analyte with the reference sensing surface to be subtracted from the response recorded for the interaction of the analyte with the target-coated surface. (Fig. 14.3).
10. We used Channel 2 as reference and checking this channel results in the correction of data. (Fig. 14.3) (see **Note 7**).
11. The "Configure button" in RI calibration can be pressed and a calibration plot is constructed when either Linear or Quadratic is selected in the pop-up window. The effect of varying solvent concentration is quantified and subtracted out of the sample signal. (Fig. 14.4).
12. The final tab is the "blanks section" (2% DMSO in running buffer), this blank injection response must be subtracted from the entire data set if this effect is present (see **Note 8**) (Fig. 14.5).
13. It is recommended to run at least three replicates of this blank and to choose the "Average option" on the Blank page (Fig. 14.5).
14. The data are now considered for final analysis which is implemented on "Kinetic page" (Fig. 14.6 and Table 14.1) (see **Note 9**).
15. In the Kinetic tab, the Two-State analysis model was selected, and "Fit" was selected. The rate equation describing the Two-State model is shown in Fig. 14.7 (see **Note 10**).

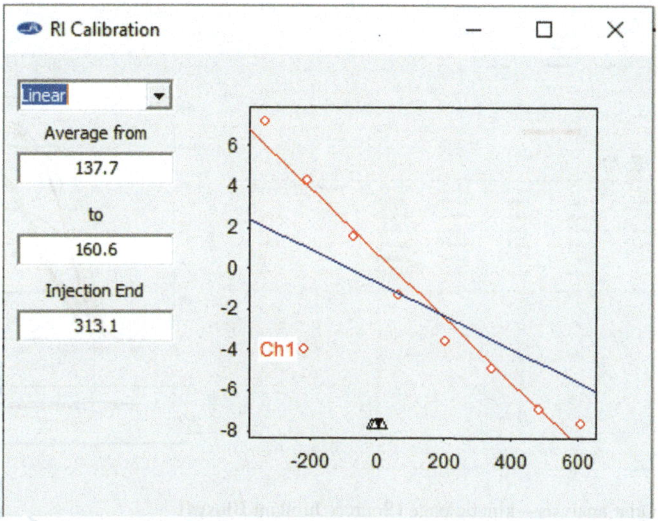

Fig. 14.4 Qdat analysis—RI calibration button (Source: Jubilant Biosys)

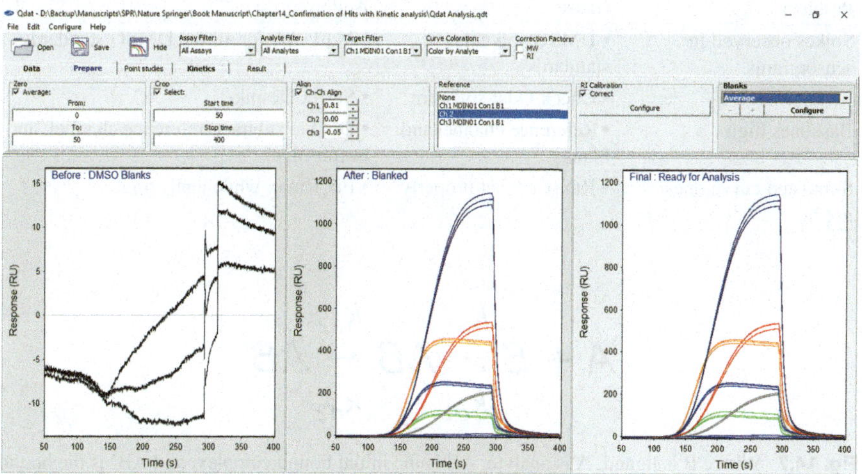

Fig. 14.5 Qdat analysis—with blank section, the data was double reference subtracted (Source: Jubilant Biosys)

16. The Fit button initiates the nonlinear regression algorithm resulting in optimization of parameter estimates (see **Note 11**).
17. Once the "fit" was completed, full binding parameters were displayed along with simulated and residual plots (Fig. 14.8) (see **Note 12**).
18. The Results tab displayed the plots generated after Qdat analysis.

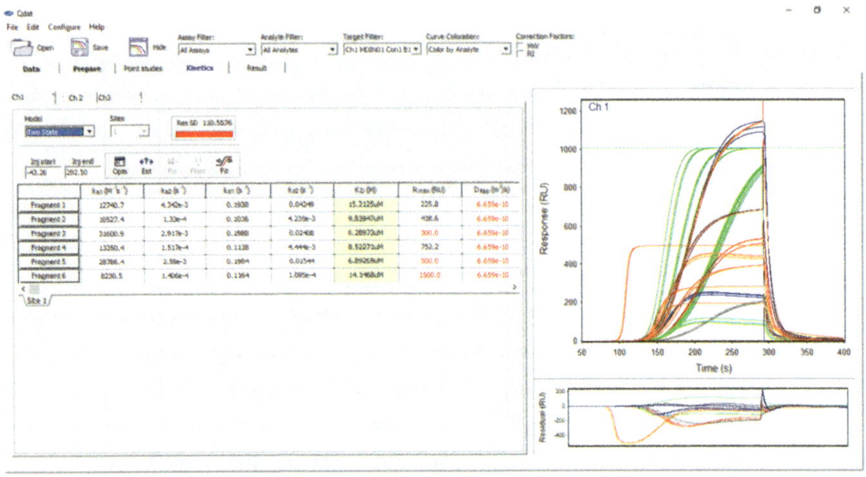

Fig. 14.6 Qdat analysis—kinetic page (Source: Jubilant Biosys)

Table 14.1 Troubleshooting guide for analysis of SPR data

Problem	Cause	Action
Spikes observed in sensorgrams	• DMSO solvent standards	• Mark "d" for all the DMSO standards
	• Check Qdat software	• Select Despike
Baselines high	• Reference channel and blanks	• Double subtract, reference channel, and buffer blank
Simulated curve does not fit	• Rmax not set properly	• Fix Rmax while analyzing

$$A + B \underset{k_{d1}}{\overset{k_{a1}}{\rightleftharpoons}} AB \underset{k_{d2}}{\overset{k_{a2}}{\rightleftharpoons}} AB'$$

Fig. 14.7 Where B is ligand, A is analyte, AB is the initial bound complex, and AB' is the bound complex in its second state. For an irreversible inhibitor which covalently binds target (Ligand), kd2 can be assumed to be 0 sec-1. Since the Two-State model is more complex, it is recommended that the user read Appendix A.3 for guidelines on when to apply a more complex kinetic model (Source: Jubilant Biosys)

19. Raw data of Un-simulated sensorgrams (Fig. 14.9) and simulated sensorgrams (Fig. 14.10) are shown here.
20. Kinetic map for hits was generated in excel sheet for better understanding of kinetic paratmers (Fig. 14.11).

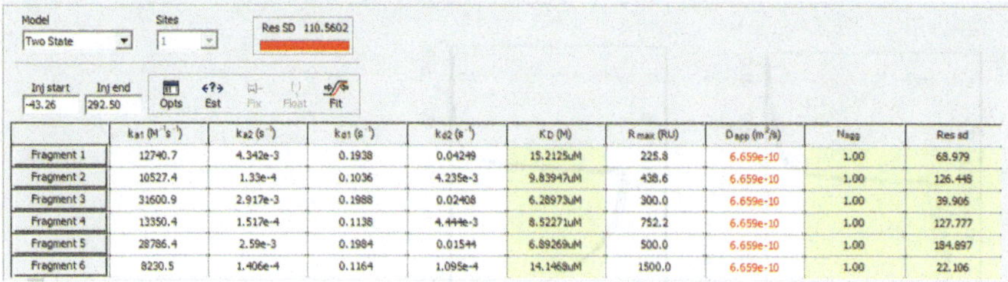

	k_{a1} ($M^{-1}s^{-1}$)	k_{a2} (s^{-1})	k_{d1} (s^{-1})	k_{d2} (s^{-1})	K_D (M)	R_{max} (RU)	D_{app} (m^2/s)	N_{agg}	Res sd
Fragment 1	12740.7	4.342e-3	0.1938	0.04249	15.2125uM	225.8	6.659e-10	1.00	68.979
Fragment 2	10527.4	1.33e-4	0.1036	4.235e-3	9.83947uM	438.6	6.659e-10	1.00	126.448
Fragment 3	31600.9	2.917e-3	0.1988	0.02408	6.28973uM	300.0	6.659e-10	1.00	39.906
Fragment 4	13350.4	1.517e-4	0.1138	4.444e-3	8.52271uM	752.2	6.659e-10	1.00	127.777
Fragment 5	28786.4	2.59e-3	0.1984	0.01544	6.89269uM	500.0	6.659e-10	1.00	184.897
Fragment 6	8230.5	1.406e-4	0.1164	1.095e-4	14.1468uM	1500.0	6.659e-10	1.00	22.106

Fig. 14.8 Qdat Analysis—full binding kinetic parameters (Source: Jubilant Biosys)

14.4 Notes

1. Preparation of fragments was explained in Sects. 11.3.1, 11.3.2, and 11.3.3.
2. File opens in Qdat analysis software—The Qdat application consists of a series of tabbed pages, each of which performs a transformation of the data set. The objective of Qdat is to rapidly perform reference curve subtraction and kinetic or affinity analyses with minimal data manipulations.
3. Specific response curves from experimental data can be viewed by selecting from one of the drop-down boxes: Assay Filter, Analyte Filter, or Target Filter. These filters default to show all data but can quickly isolate data of interest for review.
4. The color codes are predetermined by the software and the available options are: Color by Concentration, Color by Assay, Color by Analyte, and Color by Target.
5. This setting can be turned off by clicking the Configure menu button and unchecking Despike. The Number Peaks Tested defines the maximum number of spikes that can be identified and corrected in a data set. The Peak Window Size adjusts how much data is reviewed to identify and correct a spike. Larger Peak Window Sizes will extract larger spikes and lower Window Sizes will extract a more refined spike.
6. The selected data set between these limits is then shown in the right hand plot window. The data set now consists of only the regions required for model fitting (i.e., association and dissociation).
7. This transformation has completely eliminated the small response steps due to bulk refractive index offsets at the beginning and end of each injection. Any nonspecific binding or baseline drift due to other interferences are similarly subtracted, making this powerful means of isolating the responses due to analyte binding in the presence of artifacts.
8. Both a reference curve and a blank curve are subtracted from the data set before a model can be fitted. The kinetic model fit will be inaccurate if this blank buffer response is ignored. In many cases, no blank response is observed, and therefore a blank should not be subtracted from the data set as it will not improve the quality of the data set but will unnecessarily increase the random noise of the

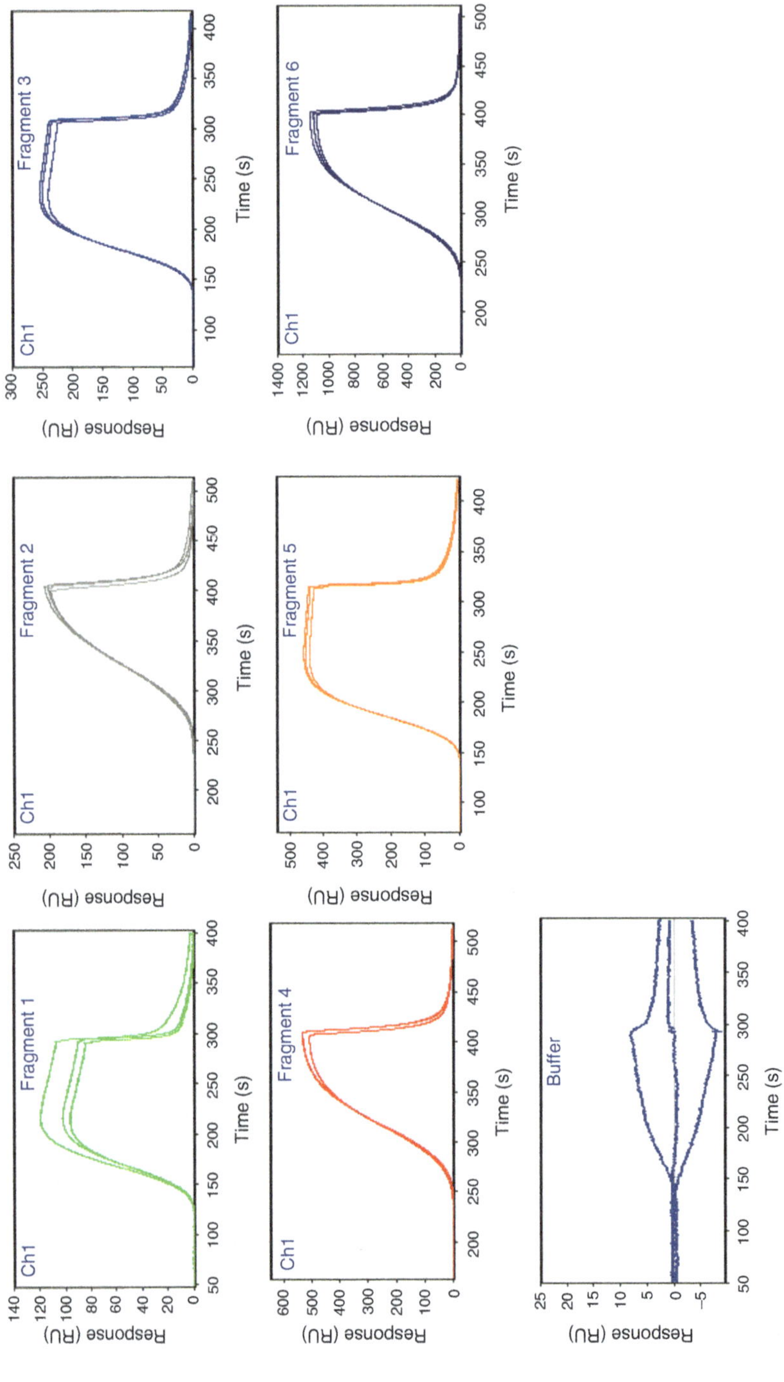

Fig. 14.9 Qdat analysis—results section—representative profiles with un-simulated sensorgrams (Source: Jubilant Biosys)

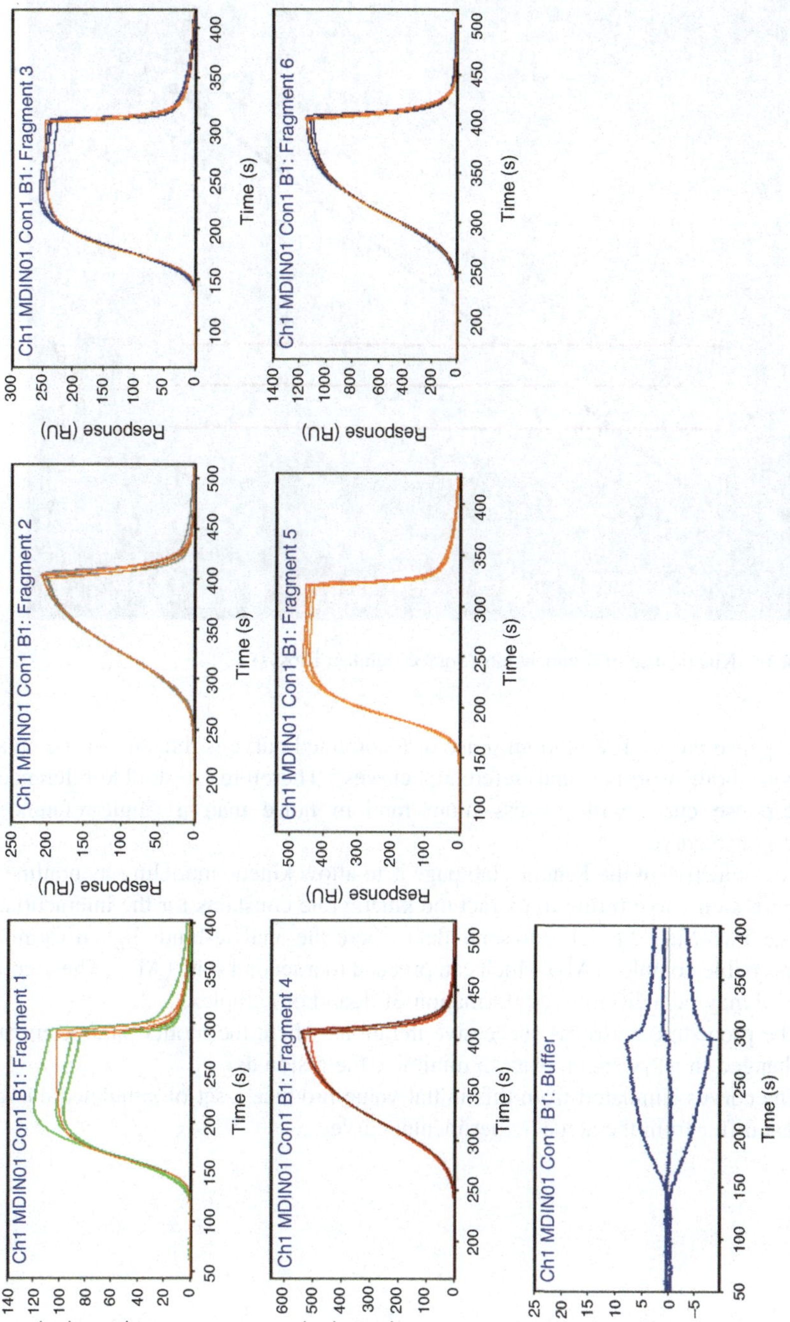

Fig. 14.10 Qdat analysis—results section—representative profiles with simulated sensorgrams (Source: Jubilant Biosys)

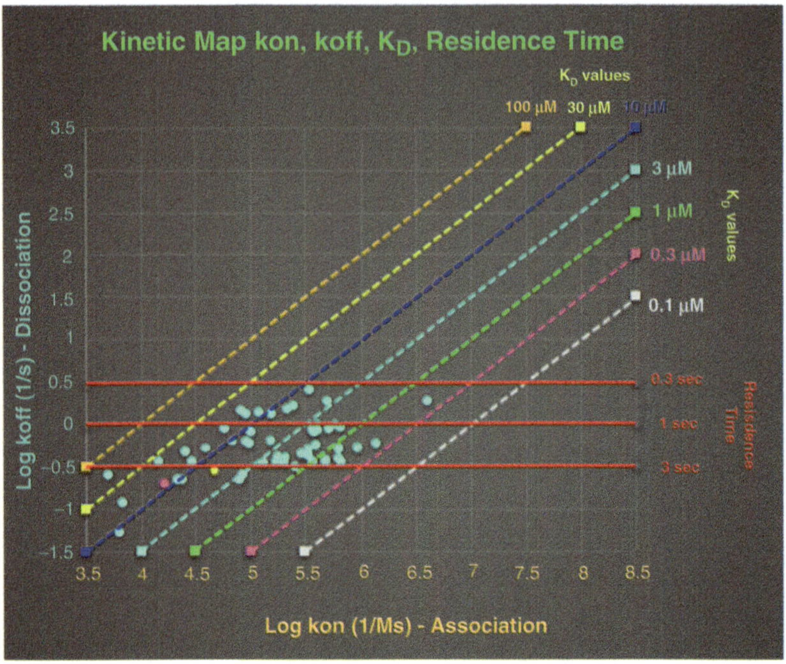

Fig. 14.11 Kinetic map of fragment hits (Source: Jubilant Biosys)

response curve. The random noise of a corrected curve is the sum of the noise from both working and reference curves: Therefore, a double-referenced response curve will possess more random noise than a single-referenced response curve.

9. The objective of the Kinetics tab page is to allow kinetic modeling by nonlinear regression curve fitting to extract the kinetic rate constants for the interaction.
10. The Two-State Model represents data where the analyte binds ligand forming reversible complex (AB) which can proceed to a second state (AB′). The second state may be a different conformation of ligand or complex.
11. The procedure performs successive iterations where the model parameters are changed in a systematic way to minimize the residuals.
12. The curves simulated from this initial value produce a set of simulated curves that differ from the actual experimental curves.

Summary

Given the challenges in identification of fragment hits in early stage drug discovery, there is a strong potential for developing SPR-based fragment screening assay. However, to be considered for using an effective screening platform, several parameters need to be examined like the fragment selection, protein with specific tag for SPR use, and proper analysis tool. Although several SPR platforms are being used for fragment screening, we provide the protocols which have been developed, optimized, and validated at each and every critical step of the screening cascade. The Pioneer FE SPR platform shows great promise in generating fragment hits that will be free of false positives. Thus in the field of preclinical in vitro drug discovery, this new platform will help in reducing time, resources to identify proper hits. With our continued efforts and rigorous assessments, we will tame the protocols for the benefit of the scientific community involved in drug discovery.

S. M. Zaheer, R. Gosu (eds.), *Methods for Fragments Screening Using Surface Plasmon Resonance*, https://doi.org/10.1007/978-981-16-1536-8

The manufacturer's authorised representative in the EU is Springer Nature Customer Service Centre GmbH, Europaplatz 3, 69115 Heidelberg, Germany. If you have any concerns regarding our products, please contact ProductSafety@springernature.com

Printed and bound by CPI Group (UK) Ltd, Croydon, CR0 4YY
30/09/2026
02225307-0004